D1186912

to l

alt wi

vs of Cork

The
CONSTRUCTION
SAFETY
HANDBOOK

T. Curley

·OAK·TREE·PRESS·

OAK TREE PRESS
19 Rutland Street, Cork, Ireland
http://www.oaktreepress.com

© 2006 T. Curley

A catalogue record of this book is
available from the British Library.

5788509

ISBN 1-904887-00-7

Printed in Ireland by ColourBooks.

CONTENTS

Chapter 3: Construction Hazards A to Z (detail)

Contents

Safety - the mission of this Handbook

FIGURES

ACKNOWLEDGEMENTS

The author wishes to acknowledge the assistance of the following organisations and individuals in the preparation of this book:

* John Curley and Claren Safety & Security Consultants Ltd.
* Construction Safety Specialists.
* Brian O'Connell Ltd.

Images and photographs are from stock photo albums or taken on-site by the author, with the permission of Brian O'Connell Ltd, Cronan Nagle Construction Ltd and Adalta Ltd, except for the following, whose use is hereby gratefully acknowledged:

* Pages 26/27 (**Safe System of Work Plan**): HSA.
* Page 65 (**Chains & Slings**): Health & Safety Executive Research Report 336 *Trojan Horse Construction Site Safety Messages*, ISBN 07176 2992 9.
* Page 130 (**Ladders**): HSA: *Code of Practice for Access & Working Scaffolds*.
* Page 186 (**Scaffolding**): HSA: *Code of Practice for Access & Working Scaffolds*.

DISCLAIMER

The author and the publisher have taken every care to ensure that the information published in this book is correct at the time of going to print. However, this book is not exhaustive in identifying hazards and risks, nor in suggesting controls and measures to eliminate or reduce them. Nor is it an interpretation of the law. Accordingly, neither the author nor the publisher can take any responsibility for any loss or damage caused to any person as a result of acting on, or refraining from acting on, any information published herein. Professional advice should be obtained before entering into any legally binding commitments.

1: Introduction

I wrote this book because, like so many other professional authors, I could not find one that covered exactly the material I wanted, in the way I wanted. As a practising Safety Adviser, responsible for a major construction site, I found that I needed regular, easy access to information about construction hazards and how to deal them. There is no shortage of information - safety is well-recognised as a priority on construction sites - but it is scattered across a multitude of sources, in a wide range of formats and usually is not accessible outside the confines of the site office. I wanted something different. Over time, my collection of notes and jottings expanded and, believing that others had the same need as myself, I resolved to compile my own book: **The Construction Safety Handbook** is the result.

1.1 How to Use the Handbook

The Construction Safety Handbook is divided into three major sections:

* **Construction Health & Safety:** A review of legislation and current best practice. This section will be of particular use to students of health and safety whose work is in a construction environment. It will also serve as a refresher for experienced construction safety professionals - not least because it incorporates up-to-date information on the Safety, Health & Welfare at Work Act, 2005.

* **Construction Hazards A to Z:** This is the core of the **Handbook**, covering nearly 100 common construction hazards. Each is dealt with in a standard format that identifies the resulting risk, suggests control measures to manage the risk, and directs the reader to legislation, standards, codes of practice, training and further information. The individual hazards are illustrated and cross-referenced, where appropriate.

* **Construction Health & Safety Directory:** A useful list of organisations that provide a wide range of information and support in the area of construction health and safety.

1.2 Glossary

Key terms used in construction health and safety include:

* **as far as reasonably practicable**
  In relation to the duties of an employer, this phrase means that an employer has exercised all due care by putting in place the necessary protective and preventative measures, having identified the hazards and assessed the risks to

safety and health likely to result in accidents or injury to health at the place of work, and where the putting in place of any further measures is grossly disproportionate, having regard to the unusual, unforeseeable and exceptional nature of any circumstances or occurrence that may result in an accident at work or injury to health at that place of work. Put simply it means that the cost (time, inconvenience or financial) of carrying out what the law requires can be balanced against the size of the risk.

* **competent person**
A person is deemed competent where, having regard to the task he or she is required to perform and taking account of the hazards, the person possesses sufficient training, experience and knowledge appropriate to the nature of the work to be undertaken.

* **dangerous occurence**
An occurrence arising from work activities in a place of work that causes or results in the collapse, overturning, failure, explosion, bursting, electrical short circuit, discharge or overload, or malfunction of any work equipment; the uncontrolled or accidental release, escape or ignition of any substance; a fire involving any substance; or any unintentional ignition or explosion of explosives. Dangerous occurrences must be reported to the HSA on IR3 forms ASAP.

* **good practice**
This refers to compliance with approved *Codes of Practice* for certain activities - for example, scaffolding, confined spaces, etc. A *Code of Practice* is not the 'law', but provides specific guidance for high risk activities.

1.3   Abbreviations

The following abbreviations are used throughout this **Handbook**:

BS	British Standard - usually followed by a number to indicate the standard referred to
CIF	Construction Industry Federation
CIS	Construction Information Sheet - a series of leaflets from HSE
CSCS	Construction Skills Certification Scheme
EN	European Norm - usually followed by a number to indicate the norm / standard referred to
EU	European Union
FÁS	*Foras Aiseann Saothair*, The National Training Authority
HSA	Health & Safety Authority
HSE	Health & Safety Executive (UK)
HSR	*Health & Safety Review* magazine
ILO	International Labour Organisation
IOSH	Institute of Occupational Health & Safety
IS	International Standard - usually followed by a number to indicate the standard referred to

MEWP	Mobile elevating working platform
MSDS	Material Safety Data Sheet
NISO	National Irish Safety Organisation
NSAI	National Standards Authority of Ireland
OEL	Occupational Exposure Limit
PPE	Personal protective equipment
PSCS	Project Supervisor Construction Stage
PSDS	Project Supervisor Design Stage
psig	Pounds per square inch gauge - a measure of pressure
RCD	Residual current device
RPE	Respiratory protective equipment
rpm	Revolutions per minute - a measure of the speed of an engine
SHWW	Safety, Health & Welfare at Work
SSWP	Safe System of Work Plan
SWL	Safe working load

You can read the **Handbook** straight-through from cover to cover, or dip into it as you need information - whatever you choose. I hope you find it useful - it certainly will make my own working day easier!

T. Curley
December 2005

Dumpers / Dump Trucks - see p. 92

2: CONSTRUCTION HEALTH & SAFETY

2.1: Irish Construction Health & Safety Legislation

The key legislation governing general construction is:

* Safety, Health & Welfare at Work Act, 2005.
* Safety, Health & Welfare at Work (General Application) Regulations, 1993.
* Safety, Health & Welfare at Work (Construction) Regulations, 2001.
* Safety, Health & Welfare at Work (Construction) (Amendment), 2003.
* Safety, Health & Welfare at Work (General Application) (Amendment) Regulations, 2001.

Proposed legislation includes:

* Safety, Health & Welfare at Work (Construction) Regulations, 2006.
* Safety, Health & Welfare at Work (General Application) Regulations, 2006.

2.1.1 Safety, Health & Welfare at Work Act, 2005

The SHWW Act, 2005 was passed by the Oireachtas and signed into law by the President in July 2005, with a commencement order of 1 September 2005.

The SHWW Act, 2005 is longer, more detailed (89 sections and 7 schedules) and more stringent than the SHWW Act, 1989, although its basic principles are the same. Its primary focus is on the prevention of workplace accidents, illnesses and dangerous occurrences. It provides also for significantly increased fines and penalties, aimed at deterring the minority who continue to flout safety and health laws.

A new provision in the 2005 Act provides that persons who commission, procure or construct a place of work must appoint a competent person to ensure that construction is carried out safely and properly and without risk to health or safety.

The other main new provisions of the SHWW Act, 2005 are:

* Fines of up to €3 million, or up to 2 years in jail, for serious breaches of health and safety regulations.
* The introduction of an on-the-spot fines system for certain safety offences.
* Drug and alcohol testing for employees.

The key parts of the SHWW Act, 2005 are:

* **Part II** sets out the duties of all persons in the workplace and is the most important part for both employers and employees. New duties are imposed, in addition to those already required by the SHWW Act, 1989:
    o Section 8: General duties of employers to their employees (see **2.2.1**).
    o Section 9: Information for employees.
    o Section 10: Instruction, training and supervision.
    o Section 11: Emergencies and serious and imminent duties.
    o Section 12: General duties of employers to persons other than their employees.
    o Section 13: General duties of employees.
    o Section 15: General duties of persons in control of places of work.
    o Section 16: General duties of designers, manufacturers, importers and suppliers.
    o Section 17: General duties of persons who carry out construction work.
* **Part III** relates to protective and preventive measures.
* **Part IV** relates to Safety Representatives & safety consultation.
* **Part V** relates to The Authority (Health & Safety).
* **Part VI** relates to regulations, *Codes of Practice* and enforcement.
* **Part VII** relates to offences and penalties.
* **Part VIII** deals with miscellaneous items.

2.1.2 Safety, Health & Welfare at Work (General Application) Regulations, 1993 - and Amendment, 2001

The General Application Regulations, 1993 and the amended Regulations, 2001, provide more explicit information than the SHWW Act, 1989 on what is required of employers to manage health and safety. These regulations consist of an amalgamation of several EU Directives, and are divided into separate parts, each dealing with a particular area:

* **Part II:** General Safety & Health Regulations.
* **Part III:** Workplace Regulations.
* **Part IV:** Work Equipment Regulations.
* **Part V:** Personal Protective Equipment Regulations.
* **Part VI:** Manual Handling Regulations.
* **Part VII:** Display Screen Equipment Regulations.
* **Part VIII:** Electricity Regulations.
* **Part IX:** First Aid Regulations.
* **Part X:** Notification of Accidents & Dangerous Occurrences Regulations.

2.1.3 Proposed Safety, Health & Welfare (General Application) Regulations 2006

The new Regulations are designed primarily to revoke and replace those provisions of the SHWW (General Applications) Regulations, 1993 that are not already incorporated into the SHWW Act, 2005.

The draft Regulations comprise 17 separate Parts and 32 associated Schedules, as follows:

* **Part I:** Interpretation & General Regulations.
* **Part II:** Workplace Regulations.
* **Part III:** Use of Work Equipment.
* **Part IV:** Provision of Personal Protective Equipment.
* **Part V:** Manual Handling of Loads.
* **Part VI:** Work with Display Screen Equipment.
* **Part VII:** Electricity Regulations.
* **Part VIII:** First Aid Regulations.
* **Part IX:** Reporting of Occupational Accidents & Dangerous Occurrences Regulations.

New Parts include:

* **Part X:** Protection of Children & Young Persons Regulations.
* **Part XI:** Protection of Pregnant, Post-Natal & Breastfeeding Employees Regulations.
* **Part XII:** Night Work & Shift Work Regulations.
* **Part XIII:** Safety Signs at Places of Work Regulations.
* **Part XIV:** Explosive Atmospheres at Places of Work Regulations.
* **Part XV:** Work at Height Regulations.
* **Part XVI:** Control of Vibration at Work Regulations.

The new Parts will revoke and replace the following existing legislation:

* Reg. 10: Safety Officers.
* Reg. 14(1)(C): Duties of employees.
* Reg. 35(1): Explosives.
* Reg. 40: Prevention of drowning.
* Reg. 41: Transport - General.
* Reg. 50: Demolition.
* Reg. 54: Working at height - defective material.
* Reg. 63: Working at height - boatswain's chairs, skips, etc. (not power-operated).
* Reg. 65: Working at height - inspection of scaffolds, boatswain's chairs, etc.
* Reg. 66: Working at height - scaffolds used by employees of different contractors.
* Reg. 69: Working at height - guard-rails & toe-boards at working platforms and workplaces.
* Reg. 70: Working at height - guard-rails, etc for gangways, runs & stairs.

* Reg.74: Working at height - openings, corners, breaks, edges & open joisting.
* Reg.79: Working at height - prevention of falls & provision of safety nets & harnesses.
* Reg.81: Lifting appliances - support, anchoring, fixing & erecting.
* Reg.83: Lifting appliances - platforms for crane drivers & signaller.
* Reg.86: Lifting appliances - brakes, controls, safety devices, etc.
* Reg.89: Lifting appliances - stability of lifting appliances.
* Reg.93: Lifting appliances - restrictions on the use of cranes.
* Reg.97: Lifting appliances - testing & examination of lifting appliances.
* Reg.100: Lifting appliances - excavators used as cranes.
* Reg.105: Chains, ropes & lifting gear - construction, testing, examinations & safe working load.
* Reg.119: Carriage of persons & secureness of loads - carrying persons by means of lifting appliances.
* Reg.123: Carriage of persons & secureness of loads - mobile elevating work platforms.
* Reg.127: Miscellaneous - keeping of records.
* Reg.128: Miscellaneous - prescribed forms.
* Reg.130: Application.

> Note: **The new proposed Regulations do not include new regulations on noise, which means that, during 2006, new Noise Regulations will be published so as to comply with Ireland's EU Treaty obligations (the Noise Directive 03/10EC).**

2.1.4 Proposed Safety, Health & Welfare (Construction) Regulations 2006

The SHWW (Construction) Regulations, 2006 are most likely to become law in early 2006. The new regulations are shorter than the SHWW (Construction) Regulations 2001, as a number of provisions have not been included in the new Regulations, including:
* **Part XIII:** Working at Heights.
* **Part XIV:** Lifting appliances.
* **Part XV:** Chains, ropes & lifting gear.
* **Part XVI:** Special provisions as to hoists.
* **Part XVII:** Carriage of persons and secureness of loads

The proposed regulations will:
* Expand the requirements of the SHWW Act, 2005 and the SHWW (General Application) Regulations, 1993.
* Give more specific direction concerning construction work.
* Revoke the SHWW (Construction) Regulations, 2001.

These proposed Regulations will apply to construction sites or engineering works, as well as places of work where construction, repair, redecoration or simple maintenance activities are being carried out. Site surveying, drilling and extraction industries are specifically excluded from the scope of these Regulations.

Changes in the new Regulations are expected to include:

* Reg. 3.1: The requirement that project supervisors be appointed in writing.
* Reg 3.3: That the Project Supervisor for the Design Stage (PSDP) be appointed at or before the start of design work.
* Reg 3.4: That the Project Supervisor for the Construction Stage (PSCS) be appointed prior to the commencement of construction.
* Reg. 3.9: Clients shall provide or arrange to have provided copies of the health and safety plan to every person being considered, or tendering, for the role of PSCS.
* Reg 3.10/3.11: Clients will be required to be reasonably satisfied that those appointed as project supervisors will devote adequate resources and have the competence to perform their duties.
* Reg 9.2: Contractors now are required to comply with directions from the PSCS (as opposed to 'take account of them', as per the 2001 Regulations).
* Reg 10: Safety officers.
* Reg 11: Duties of employees.
* Reg 40: Prevention of drowning.
* Reg 41: Transport.
* Reg 50: Demolition.
* Reg 54: Record-keeping.

2.1.5 Other Relevant Legislation

Other legislation relevant to the construction industry includes:

* European Communities (Protection of Workers) Exposure to Noise Regulations, 1990.
* SHWW (Signs) Regulations, 1995.
* SHWW (Night Work & Shift Work) Regulations, 2000 .
* European Communities (Protection of Workers) Exposure to Asbestos Regulations, 1989-2001.
* SHWW (Confined Spaces) Regulations, 2001.

> Note: **The proposed SHWW (General Application) Regulations, 2006, are intended to replace the Signs Regulations, 1995 and the Night Work & Shift Work Regulations, 2000.**

Legislation relating to the use of chemicals includes:

* SHWW (Biological Agents) Regulations, 1994 & 1998.
* SHWW (Chemical Agents) Regulations, 2001.
* SHWW (Carcinogens) Regulations, 2001.

2.1.6 Enforcement of Health & Safety Legislation

Source: SHWW Act, 2005, s.62.

The HSA's Inspectorate enforces legislation on occupational safety, health and welfare.

Powers of HSA Inspectors (SHWW Act, 2005, s.64)
In addition to rights of entry, inspection and sampling, an inspector may:

* Inspect, remove and retain records.
* Issue an Improvement Direction, to which an employer is required to respond with an Improvement Plan in relation to activities that the inspector considers may involve risk to safety or health of persons.
* Issue an Improvement Notice stating his opinion that an employer has broken a provision of an Act or Regulation.
* Issue a Prohibition Notice in relation to an activity that the inspector is of the opinion has been, or is likely to be, a risk of serious personal injury to persons at work. This might require an immediate stoppage of work.
* In certain cases, recommend the initiation of prosecutions.
* Give directions or instructions.

It should be noted that the list above is not exhaustive. For a full list of inspectors' powers, see SHWW Act, 2005, s64.

If an inspector is prevented from entering any workplace, he / she may apply to the District Court for a warrant.

High Court Orders (SHWW Act, 2005, s.71)
In certain circumstances, the HSA may apply under section 71 of the 2005 Act to the High Court for an Order in relation to certain activities. This can be obtained on an *ex parte* basis - that is, without notice to the employer concerned. Such Orders are usually obtained where physical hazards and significant deficiencies in health and safety management cause the HSA to consider the risk to persons to be very serious.

Improvement Directions & Plans (SHWW Act, 2005, s.65)
If an inspector, while inspecting an workplace, finds a risk that is not too serious but needs to be looked at, he / she will issue the person in charge of the workplace with an Improvement Direction.

This requires the employer to prepare and submit to the HSA a plan of action (an Improvement Plan), which specifies what remedial action will be taken to reduce or eliminate the risk and a timeframe within which the action will be taken.

If an employer ignores an Improvement Direction, fails to produce an Improvement Plan or fails to implement the agreed plan, the inspector may issue an Improvement Notice.

Improvement Notices (SHWW Act, 2005, s.66)
More serious offences are dealt with by Improvement Notices.

An Improvement Notice is issued when, in the opinion of the HSA inspector, the

employer is contravening legislation. An
Improvement Notice is a direction to an
employer that states what action is required
and by when, although the compliance date
may be changed at the discretion of the
HSA.

Prohibition Notices (SHWW Act, 2005, s.67)

If an inspector believes there is an
imminent risk of death or serious personal
injury - for example, in relation to work at
heights / structural collapse - he / she may
issue a Prohibition Notice, which requires
immediate action. The inspector can
suspend all work activities, except those
required to ensure the immediate safety of
persons.

The HSA can apply to the High Court to
obtain an Order to close a place of work
until certain conditions are met, in
situations where it believes that the
continued use of the place of work presents a serious risk of death or serious injury
to employees. This Order is similar to a Prohibition Notice, except that there is no
right to appeal.

Fines (SHWW Act, 2005, s.77 & 78)

Section 77 of the SHWW Act, 2005 provides for two categories of offences: the first
applies to less serious matters, while the second covers all of the more serious
offences under health and safety laws.

Section 78 of the SHWW Act, 2005 provides for a fine not exceeding €3,000, under
summary conviction, for a person guilty of an offence under the first category of
offences set out in section 77 (less serious offences). A person guilty of any other
offence set out in section 77 is liable, on summary conviction, to a fine not exceeding
€3,000 or imprisonment up to 6 months, or both. On conviction on indictment for a
more serious offence, the maximum fine is €3 million or imprisonment for up to 2
years, or both.

On-the-spot fines (SHWW Act, 2005, s.79)

A system of on-the-spot fines by inspectors has been introduced in the SHWW Act,
2005, which provides that the level of on-the-spot fine must not exceed €1,000.

The HSA will not initiate a prosecution before the due date of payment of an on-
the-spot fine and, if the payment is made in time, no prosecution will be launched. If
a prosecution is taken, the onus is on the accused to prove that payment has been
made.

2.2: Duties & Responsibilities

2.2.1 General Duties of Employers

Source: SHWW Act, 2005, s.8; SHWW (General Application) Regulations 1993, s.5.

> **Note: SHWW (General Application) Regulations, 1993, Regs.17, 19, 21, 28, 31, 35 & 56, as well as other legislation - for example, Noise Regulations - also confer specific duties on employers.**

Current legislation, both European and Irish, imposes on employers a range of duties and responsibilities - to manage and conduct all work activities so as to prevent improper conduct or behaviour likely to put the health and safety of employees at risk. Specifically, these include:

* To **employees**:
  o To provide a properly designed and maintained place of work that is safe and without risk to health.
  o To provide a properly designed and maintained means of egress and access to and from the place of work.
  o To provide properly designed and maintained plant and machinery that is safe and without risk to health.
  o To provide a properly planned, organised, performed and maintained system of work that is safe and without risk to health.
  o To provide such information, instruction, training and supervision as is necessary for ensuring the safety and health of their employees.
  o To provide suitable PPE, free of charge to employees, where the use of such equipment is exclusive to the place of work.
  o To prepare and revise adequate plans to be followed in emergencies.
  o To provide and maintain facilities to ensure the welfare of their employees.
  o To obtain the services of a competent person for the purpose of ensuring the safety and health of employees.
  o To ensure that any measures taken relating to health and safety do not involve a financial cost to an employee.
  o To prepare a Safety Statement and to bring it to the attention of all employees, and any other person who may be affected - for example, visitors, etc.
  o To carry out Risk Assessments and to keep a written copy of those that pose a significant risk.
  o To review Risk Assessments when there has been significant change to the work processes, equipment, etc or there is reason to believe the current Risk Assessment is no longer valid.
  o To provide information to employees on matters of safety and health.
  o To consult with employees on matters of safety and health.
  o Provide training on matters of safety and health, without loss of remuneration to employees.

o To ensure health surveillance is made available for every employee, appropriate to the health and safety risks.

o To prevent a risk to health arising from the use of any substance or article or exposure to noise, vibration or ionising or other radiation or any other physical agent.

o To provide adequate instruction, training and supervision to all employees at the time of recruitment in a form, manner and language that employees are reasonably likely to understand.

o To appoint competent persons to perform the functions relating to protecting the employees from risks to their health and safety.

* To **non-employees** (**Source:** SHWW Act, 2005, s.12):

o To conduct all business activities in such a manner that the health and safety of persons not under direct employment are not put at risk.

o To inform all persons of any activity that may affect their health and safety.

A banksman - see p.136

2.2.2 General Duties of Employees

Source: SHWW Act, 2005, s.13; SHWW (General Application) Regulations, 1993, s.14; SHWW (Construction) Regulations, 2001, s.14.

An 'employee' is defined by the 2005 Act as meaning 'a person who has entered into or works under a contract of employment' and includes a fixed term employee and temporary employee.

Under the Regulations, employees have legal responsibilities to:

* Take reasonable care of their own safety and health and that of others who may be affected by their actions.

* Co-operate with management to meet the employer's legal duties.

* Use any device or protective equipment intended to help secure their safety or health.

* Make correct use of machinery, tools, dangerous substances, transport equipment, etc.
* Report to management any defects in equipment or other dangers immediately, or as soon as it is safe to do so.
* Not to intentionally or recklessly interfere with or misuse anything provided in the interest of health, safety or welfare.
* Comply with site rules, as contained in the Safety & Health Plan (see **2.3.2**).
* Make proper use of safety equipment, such as safety helmet, harness and footwear.
* Undertake training in relation to the FÁS Safe Pass programme and Construction Skills Certification Scheme (CSCS) (and any other training in relation to health and safety).
* Produce Safe Pass and relevant CSCS cards when requested to do so by the Project Supervisor Construction Stage (PSCS), their employer or by a HSA inspector.
* Not to be under the influence of an intoxicant to the extent that they may endanger their own safety or that of others.
* To attend training and undergo such assessment as may be reasonably required.

> Note: **An employee, on entering a contract of employment, must not misrepresent the level of training that he / she has received previously.**

2.2.3 General Duties of Clients

Source: SHWW (Construction) Regulations, 2001, s.3; *proposed SHWW (Construction) Regulations, 2006.*

According to the Regulations, a 'client' means 'any person, engaged in trade, business or other undertaking, who commissions or procures the carrying out of a project or who undertakes a project directly for the purpose of such trade, business or undertaking'.

The client is the person arranging for work to be carried out and, in some situations, carrying out the work themselves. The client may set the tone for the project by making crucial decisions regarding how the contracts are set up, the budget and timescale restraints and the selection of the designers and contractors who will carry out the work.

Clients have the following duties:

* To make appropriate notifications to the HSA.
* To appoint, in writing, a Project Supervisor for the Design Stage (PSDS) and a Project Supervisor for the Construction Stage (PSCS).
* To be reasonably satisfied that the persons appointed to those roles are competent and have adequate resources.
* To co-operate with the persons appointed to the above roles.
* To allow the PSDS and PSCS sufficient time to carry out their functions and

not make unreasonable demands regarding timescales.

* To provide relevant information about existing structures.
* To ensure that a copy of the preliminary Safety & Health Plan is provided to all those tendering for the role of PSCS.
* To ensure that construction work does not start until the Safety & Health Plan has been prepared.
* To retain the Safety File for the project and pass it on to any new owner.

> **Note: Just as other technical enquiries would be made by a client about a potential designer or contractor, safety must be a key aspect of the enquiries made by a client with regard to potential appointees before an appointment is made. This might include enquiries regarding training, qualifications and experience of staff who will work on the project, references from other clients, resources to be devoted to the project and resources that can be called upon.**

There is nothing to prevent the client appointing themselves as project supervisor, if competent to do so, or appointing one person for both roles of PSDS and PSCS.

The duties of a client relate only to projects carried out in relation to trade or business and do not apply to persons appointing a building contractor to carry out construction work such as building, or extending, a home.

2.2.4 General Duties of the Project Supervisor Design Process (PSDS)

Source: SHWW (Construction) Regulations, 2001, s.4; *proposed SHWW (Construction) Regulations, 2006.*

The PSDS is the organisation or individual appointed by the client in writing to carry out the duties and responsibilities of the PSDS for the duration of the project. This appointment may be renewed or changed during the duration of the project by the client, as required.

The PSDS is responsible for the co-ordination of all designers involved in the project and the coordination of the activities of other persons engaged in work related to the design. Other duties of the PSDS include:

* To take account of the general principles of prevention (see **Figure 1**), any existing Safety & Health Plan (see **2.3.2**) or Safety File (see **2.3.5**) in the design, phasing and workload estimating of the project.
* To prepare the preliminary Safety & Health Plan.

* To supply the required information to the PSCS: a general description of the project, its intended schedule / phasing, and work that involves a particular risk.
* To supply information required to create the Safety File.

The responsibilities of the PSDS may be delegated to, or facilitated by, the appointment of a Health & Safety Co-ordinator for the Design Stage.

2.2.5 General Duties of Designers
Source: SHWW Act, 2005, s.16; SHWW (Construction) Regulations, 2001, s.6.

Designers are organisations or individuals who undertake design work for a project, including the design of temporary works. They are in a unique position, and often can make decisions that can reduce significantly the risks to construction health and safety. Recent developments in legislation have recognised the importance of the role designers can play in the overall health and safety of the construction of a project.
    Designers include:
* Architects, civil and structural engineers, building surveyors, landscape architects and other design practices and individuals who contribute to, or have a responsibility for, analysing, calculating, preparatory design work, designing, detailing, specifying and / or preparing bills of quantities for construction work.
* Mechanical, Electrical, Chemical and other engineers.

Figure 1: General Principles of Prevention

(a) The avoidance of risks.

(b) The evaluation of unavoidable risks.

(c) The combating of risks at source.

(d) The adaptation of work to the individual, especially as regards the design of places of work, the choice of work equipment and the choice of systems of work, with a view, in particular, to alleviating monotonous work and work at a predetermined work rate and to reducing their effect on health.

(e) The adaptation of the place of work to technical progress.

(f) The replacement of dangerous articles, substances or systems of work by non dangerous or less dangerous articles, substances or systems of work.

(g) The development of an adequate prevention policy in relation to safety, health and welfare at work, which takes account of technology, organisation of work, working conditions, social factors and the influence of factors related to the working environment.

(h) The giving of priority to collective protective measures over individual protective measures.

(i) The giving of appropriate training and instruction to employees.

Source: SHWW (General Application) Regulations, 1993, First Schedule, Reg.5.

* Those who specify or modify a design, or who specify the implementation of certain methods of work or the use of specific materials (this may include the client).

Those employing, or in control of, people undertaking design work are themselves deemed to be designers.

Design work might be classified as:

* Permanent works design by the client's designers.
* Permanent works design by designers employed by specialist sub-contractors / suppliers.
* Temporary works design by designers employed by the contractor and, in some cases, by the specialist sub-contractors / suppliers.

> Note: **The Regulations identify different classifications of designers.**

Under the Regulations, designers have the following duties:

* To take account of the general principles of prevention (see **Figure 1**), during the design and when estimating the period of time required for the completion of the project.
* To co-operate with, and take into account any reasonable directions given by, the PSDS or PSCS.
* To provide information to the PSDS or PSCS relating to the risks to the safety and health of persons required to work on the project.
* To make available appropriate information on health and safety aspects of their designs for the benefit of contractors and the other designers, and to provide the information necessary for the contractor to identify and to manage the risk.
* To provide information on the residual risks of the resulting structure for inclusion in the Safety File.

All designers should carry out written Design Risk Assessments at the various stages of design development. Risk Assessments are essential at a number of specific stages in the life cycle of every project and are site-specific, not generic in nature.

2.2.6 General Duties of the Project Supervisor Construction Stage (PSCS)

Source: SHWW (Construction) Regulations, 2001, s.6.

The PSCS is the organisation or individual appointed by the client in writing to carry out the duties and responsibilities of the PSCS for the duration of the project. This appointment may be renewed or changed during the duration of the project by the client, as required.

The PSCS is ultimately responsible for managing and co-ordinating construction phase safety and health issues on site.

The duties of the PSCS are to:

* Notify the HSA of all notifiable projects.
* Co-ordinate the application of the general principles of prevention (**Figure 2**).
* Develop the Safety & Health Plan, as supplied by the PSDS, prior to construction work starting.
* Update and review the Safety & Health Plan, as required, during construction work.
* Develop and update the Safety File, as prepared by PSDS, and pass it over to the client on completion of the project.
* Co-ordinate the safety and health activities of all contractors throughout the duration of the construction period.
* Take reasonable measures to ensure that no unauthorised person enters the work area.
* Facilitate the putting in place of a site Safety Representative, where there are more than 20 people working on site.
* Appoint in writing a full-time Safety Officer, where more than 100 persons on site at any one time are engaged in construction work.
* Co-ordinate arrangements for the provision and maintenance of welfare facilities for all persons at work on a construction site.
* Co-ordinate the notification of accidents and dangerous occurrences and keep a record of same on site for inspection.
* Co-ordinate the checking of valid certificates / cards relating to the Safe Pass programme and CSCS.

Figure 2: Application of General Principles of Prevention to Construction

(a) Keeping the construction site in good order and in a satisfactory state of cleanliness.

(b) Choosing the location of workstations, bearing in mind how access to these workplaces is obtained and determining routes or areas for the passage and movement of equipment.

(c) The conditions under which various materials are handled.

(d) Technical maintenance, pre-commissioning checks and regular checks on installations and equipment with a view to correcting any faults which might affect the safety and health of persons at work.

(e) Co-operation between employers and self-employed persons; the demarcation and laying-out of areas for the storage of various materials, in particular where dangerous materials or substances are concerned.

(f) The conditions under which the dangerous materials used are removed.

(g) The storage and disposal or removal of waste and debris.

(h) The adaptation, based on progress made with the site, of the actual period to be allocated for the various types of work or work stages.

(i) Interaction with industrial activities at the place within which, or in the vicinity of which, the construction site is located.

Source: SHWW (Construction) Regulations, 2001, Third Schedule, Regs.6 & 9.

* Co-ordinate and facilitate safety consultation with employees.
* Ensure that all contractors comply with any safety requirements applicable to them as laid out in the Safety & Health Plan.

2.2.7 General Duties of Contractors

Source: SHWW (Construction) Regulations, 2001, s.9.

Contractors are those organisations or individuals who actually carry out the construction work on the project. Depending on the size of the project, there will be a main contractor (normally nominated as the PSCS), who in turn will appoint sub-contractors, who in turn will appoint further sub-contractors. Depending also on the size of a project, a sub-contractor's job can last from three months to three years. Block-layers are in demand early in a project, followed by roofers, carpenters, plasterers, electricians, plumbers, painters, and decorators.

According to the Regulations, a 'contractor' means 'a contractor or an employer, whose employees undertake, carry out or manage construction work, or any person who carries out or manages construction work for a fixed or other sum and who supplies the material and labour (whether his own labour or that of another) to carry out such work or supplies the labour only'.

The duties of a contractor are to:

* Satisfy themselves that any contractors or designers they engage are competent and adequately resourced to perform their roles.
* Provide a copy of their Safety Statement to the PSCS.
* Comply with any reasonable directions from the PSCS and with any relevant provisions in the Safety & Health Plan.
* Inform the PSCS of all accidents and dangerous occurrences.
* Provide information for the Safety File (see **2.3.5**).
* Provide information and training to their employees.
* Ensure consultation with the workers.
* Ensure that all their employees who carry out construction work are in possession of Safe Pass certification.
* Ensure that workers carrying out any tasks as listed in Schedule 9 of the Regulations are in possession of the appropriate CSCS cards.
* Appoint a competent Safety Officer to advise on, and supervise adherence to, Health & Safety requirements, if there are normally more than 20 employees on a site, or more than 30 engaged in construction on various sites.
* Ensure that employees are adequately trained in their trade (for example, a trade apprenticeship plastering, carpentry, electrical works, etc).
* Ensure compliance with the technical requirements to do with excavations, scaffolding, working at heights and other specific elements of the Construction Regulations.

As an employer, the contractor must also comply with other statutory provisions as laid out in the SHWW Act, 1989 and associated legislation.

2.3: Safety Documentation

2.3.1 Safety Statement

Source: SHWW Act 2005, s.20.

Why do I need a Safety Statement?

It is a legal requirement that all employers, including the self-employed, must have a written Safety Statement, which must be brought to the attention of all employees or other persons who may be affected by its contents.

Not having a Safety Statement is an offence and can lead to prosecution by the HSA. Apart from the threat of legal action, a written Safety Statement provides commitment from management towards health and safety. It also gives clear instruction to all employees as to who is responsible for what and the measures that should be taken to minimise the risk and to implement control measures, for everyone's safety.

Who should prepare the Safety Statement?

The Safety Statement should be prepared by a competent person who understands the activities of the company (for example, safety advisor, project manager, etc.) and the approved document must be signed by senior management, ideally by the Managing Director.

Safety Statements may be prepared by an external safety consultant, although they can be prepared by a director / owner of the company. What is important is that the person preparing the statement is competent to do so and has the relevant background experience in the industry.

A Safety Statement must be relevant and specific to each business. It is not good enough to take another company's statement and just change the details.

The Managing Director is responsible for appointing a competent person to prepare the Safety Statement for the company.

What exactly is in a safety statement?

A Safety Statement shows a commitment by company management to run the business in a safe manner and states both the means and resources by which this will be accomplished, who will carry out those activities and the resources which are required / available to achieve a safe working environment

A good Safety Statement will help to display legal compliance. It will refer to almost every aspect of the employer's business and will be current and updated regularly.

The HSA has identified three components of a good Safety Statement:

* **Policy Statement:** This is the company statement of its commitment to health and safety. It includes a declaration that the company will manage all its activities without disregard for the safety of employees, visitors, clients, sub-contractors, members of the public, etc. The reduction / elimination of accidents and the protection of health of all persons affected should be explicitly stated as a high priority and a key feature of the company's activities.

* **Organisation:** Who has certain duties and who is responsible for what? This section should detail the company management and organisation,

2.3.5 Safety File

The 'Safety File' is a record of information on the key health and safety risks that have to be managed during future maintenance, repair or construction work, following completion of the construction project. The information it contains will alert those who are responsible for the building, and services in it, of the important safety and health risks that will need to be considered during subsequent maintenance, repair or refurbishment, extension or other construction work.

> **Note: The Safety File is often referred to as an 'Operation & Maintenance Manual', as it contains operation and maintenance manuals of all plant within the building, as-built or as-installed drawings and any other information required to ensure the safety of people working on the building in future.**

Who is responsible for preparing the Safety File?

Under Regulation 4(1)(b), the PSDS has responsibility for preparing the Safety File and, upon completion of the project, for delivering it to the client. If a client subsequently disposes of his / her interest in the property, the Safety File must be passed on from the client to the person who is now responsible for the structure.

Ideally, the file should be prepared on an ongoing basis throughout the project by the PSCS and handed over to the client on completion of the project. In some cases, it might not be possible for a fully-developed Safety File to be handed over at the end of construction of the project. This may happen because the construction work had to be finished rapidly to meet a tight deadline and completion of the Safety File was impossible while work was ongoing. Clearly, a commonsense approach is needed, allowing the Safety File to be handed over as soon as practical after the completion certificate (practical completion) or similar document has been issued. What is important is that work on producing the file continues throughout the project and is not left until the end.

What does a Safety File contain?

Safety Files should be in proportion to the scale of the project. Smaller projects will require straightforward files, while large structures with significant risks will need more detail.

Information recommended by the HSA for the Safety File includes:

* All 'as-built' drawings, final plans, specifications and bills of quantities, used and produced throughout the construction process.
* The general design criteria.
* Details of the equipment and maintenance facilities within the structure.
* Maintenance procedures and requirements for the structure.
* Manuals, and where appropriate, certificates, produced by specialist contractors and suppliers, which outline operating and maintenance procedures and schedules for plant and equipment installed as part of the structure - typically, lifts, electrical and mechanical installations and window-cleaning.
* Details of the location and nature of utilities and services, including emergency and fire-fighting systems.

Risk Ratings

A simple formula can be used to determine the significant risk rating posed by hazards (Clarke):

Noise & Vibration - see p.155/216

$$\text{Risk Rating} = \text{Consequence x Likelihood}$$

where:

* ✶ Risk Rating 1 - 9 = Low
* ✶ Risk Rating 10 - 15 = Medium
* ✶ Risk Rating 16 - 25 = High

Consequence categories

Category	Value	Description
Major	5	Causing death to one or more people. Loss or damage is such that it could cause serious business disruption - for example, structural damage, fire.
High	4	Causing permanent disability (loss of limb, sight, hearing) and / or death.
Medium	3	Causing temporary disability (fractures) or damage.
Low	2	Causing significant injuries (sprains, lacerations).
Minor	1	Causing minor injuries (cuts, scratches). No lost time likely, other than for first aid treatment.

Likelihood categories

Category	Value	Description
Almost Certain	5	Absence of any management controls. If conditions remain unchanged, there is almost 100% certainty that an accident will happen.
High	4	Serious failures in management controls. The effects of human behaviour or other factors could cause an accident, but it is unlikely without this additional factor (ladder not secured properly, poorly trained personnel, etc).
Medium	3	Insufficient controls in place. Loss is unlikely during normal operation, but may occur in emergencies or non-routine conditions.
Low	2	The situation is generally well-managed, although occasional lapses could occur. This also applies to situations where people are required to behave safely in order to protect themselves but are well-trained.
Improbable	1	Loss, accident or illness could only occur under freak conditions. The situation is well-managed and all reasonable precautions have been taken.

For example, take working at height. The consequence is High = 4; a fall could cause permanent disability and / or death. The likelihood (where no controls are in place) is Almost Certain = 5. The Risk Rating then is:

$$\text{Risk Rating} = \text{Consequence} \times \text{Likelihood}$$
$$4 \times 5 = 20$$
$$\text{Risk Rating} = \text{High (16 - 25)}$$

This system produces numeric values for the risk. Although simplistic, this is a useful and fast method for determining the risks that are considered significant.

The results of all Risk Assessments should be documented in writing (regardless of whether the outcome was high / medium / low). Protective measures and **Controls → Managing the Risks** should be detailed, and this information included in safety documentation such as the Safety Statement / Safety & Health Plan.

All Risk Assessments should to be reviewed periodically to ensure that the **Controls → Managing the Risks** are still adequate to the risk.

2.4.2 Safe System of Work Plan

The Safe System of Work Plan (SSWP) is a new initiative, launched by the HSA, aimed at reducing injuries and deaths on construction sites.

It relies heavily on pictograms to explain and clarify hazards and controls, thereby creating a wordless document through which safety can be communicated to all workers, regardless of literacy or language skills. The SSWP focuses on those in the construction industry who are most at risk and empowers them to ensure that all necessary safety controls are in place prior to the commencement of planned work.

How do I use the SSWP?
The process consists of 3 parts:
1. Planning the activity.
2. Identifying the hazards and controls.
3. Sign-off.

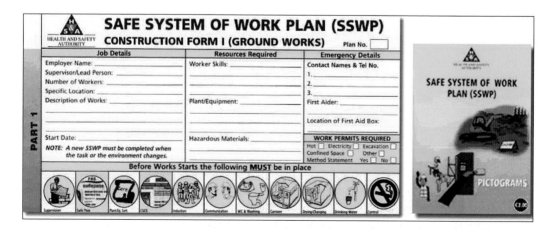

Planning the activity

The person planning the activity will complete this part, prior to work starting. Normally, this would be the supervisor/foreman and/or self-employed person. Where a site safety officer is employed, they should be involved in the process.

Identifying the hazards and controls

This part of the SSWP form deals with hazard identification, risk assessment and risk control. Normally this would be carried out by the supervisor/foreman and/or self-employed person, prior to work starting. Where a site safety officer is employed, they should be involved in the process.

The hazards should first be identified by ticking the square boxes in the 'Select Hazard' column.

✴ The appropriate Controls to eliminate the hazard or reduce the risk should be identified by ticking the square boxes in the "Select Control" column

✴ When the Controls are in place tick the round box. This must be done in conjunction with the workers at the specific work location prior to the work taking place

✴ The Personal Protective Equipment (PPE) and Fire Equipment required should be selected by ticking the square boxes in the PPE/Fire sections and when acquired by ticking the round box

Sign off

The purpose of signing-off is to identify the person who has prepared the SSWP and also to confirm that the completed SSWP has been brought to the attention of all those to whom the SSWP applies.

PART 3

SSWP prepared by: _____ **Date:** _____

The controls to be used as per this form have been brought to my attention.

Signed by Team:

NOTE: This list of Hazards and Controls is not exhaustive and is in no particular order.

IF IT'S NOT SAFE DON'T DO IT AND INFORM SITE MANAGEMENT

Note: **The completed SSWP must remain with at the specific location of the work with the persons carrying out the work activity. A new SSWP must be completed when (1) a new hazard is identified, (2) the task changes, or (3) the environment changes.**

The SSWP is currently available from the HSA for groundworks and house-building. The HSA has also published Turkish and Polish versions of the SWPP (see **www.hsa.ie**).

2.4.2 Site Inspections

Formal inspections of the construction site, at reasonably regular intervals, are a mechanism for augmenting day-to-day checks, examinations or inspections that occur as part of any task. Many Safety Officers / Safety Representatives use a checklist to facilitate consistency and to allow for the easy identification of issues. (Sample site inspection checklists are available on **www.oaktreepress.com**.)

The advantages of a regular inspection regime are that it assists in the consistent maintenance of workplace conditions, highlights problem areas and helps raise awareness among all employees of the need to maintain and improve safety standards. It may also assist in highlighting safety offenders.

Disadvantages include that inspections can become purely routine, or are not acted upon. Site inspections should be carried out without prior notice, at different times and on different days of the week, to avoid predictability.

In practice, a combination of both regular and random inspections is advisable.

Remedial actions

There is little point in identifying areas of concern, if nothing is done about them. A clear understanding of what remedial actions are required is fundamental to safety inspections. Risks must be removed, or reduced to an acceptable level, and the workplace made safer. A plan of what needs to be done by whom and by when may be required. Identified risks and remedial actions to be put in place should be discussed at site safety meetings / site management meetings / contractors meetings and toolbox talks. The results need to be brought to everyone's attention.

The results of inspections may be categorised as:

* Situations likely to result in imminent fatality or serious injury to the public or workers, if immediate action is not taken - immediate action.
* Hazards exist and prompt action is required, but no imminent danger - prompt action (within 24 hours)
* Possibility of minor loss or injury - remedial action within 7 days.

However, if an issue arises repeatedly, the reason for the reoccurrence needs to be identified and controls put in place to prevent it happening at all.

2.5: Training

According to the HSA, "Training has long been seen as the 'delivery system' for health and safety". Employers are required to provide information, instruction, training and supervision to employees, so far as reasonably practicable, under the following legislation:

* SHWW Act, 2005, s.2; s.8; s.10; s.11; s.18; s.25; s.77; s.78; Schedules 3 to 7.
* *Proposed SHWW (General Application) Regulations, 2006, Part IV, Reg.13; Part*

VI, Reg.18; Part VIII, Reg.41; Part X, Reg.51; Part XIII, Reg.65; Part XIV, Schedule 22.7.

* *Proposed SHWW (Construction) Regulations, 2006, Reg.6; Reg.7; Reg.8; Reg.9; Reg.14; Reg.56; Schedules 5 and 6.*
* SHWW (General Application) Regulations, 1993, Reg.13.
* SHWW (Construction) Regulations, 2001, Part III.

Employees must be trained on commencing employment and/or entering onto a new site. All training, instruction and supervision must be in a form, manner and language that employees are reasonably likely to understand. All training, instruction and information should be task-specific and cover emergency measures.

As well as training on commencement, employees should receive training if they are transferred or assigned to other tasks, and on the introduction of new work equipment / technology and systems of work.

Numerous studies have shown that correct training reduces the likelihood of accidents among construction workers.

> **Note:** **Employee training should be provided during normal working hours and without loss of remuneration.**

2.5.1 Safe Pass Programme

Source: *Proposed SHWW (Construction) Regulations, 2006, Reg.6; Reg.9; Reg.14; Reg.56; Schedule 5.*

Under the Regulations, all persons working on a construction site must have completed the basic FÁS Safe Pass safety awareness programme, which became mandatory on 1 June 2003. Following a review of the entire programme, Safe Pass Phase II commenced in July 2005, introducing new graphics and safety pictures, the use of video and guidance of the use of safe working procedures.

On completion of the Safe Pass programme, workers receive a registration card, which must be produced on request on any construction site.

The Safe Pass programme consists of 12 modules:

* Promoting a safety culture.
* Health and safety legislation.
* Accident reporting and emergency procedures.
* Working at heights.
* Excavations and confined spaces.
* Electricity - underground and overhead services.
* Personal protective equipment.
* Use of hand-held equipment, tools and machinery.
* Safe use of vehicles.
* Noise and vibrations.
* Manual handling.
* Health and hygiene.

2.5.2 Construction Skills Certificate Scheme (CSCS)

Source: *Proposed SHWW (Construction) Regulations, 2006, Reg.6; Reg.9; Reg.14; Schedule 6.*

Under the Regulations, workers carrying out safety-critical tasks are required to complete a training course and be certified.

On completion of the CSCS, workers receive a registration card, which must be produced on request on any construction site.

The safety-critical tasks requiring CSCS certification are:

* Scaffolding - basic.
* Scaffolding - advanced.
* Tower crane operation.
* Slinging / Signalling.
* Telescopic Handler operation.
* Tractor / Dozer operation.
* Mobile Crane operation.
* Crawler Crane operation.
* Articulated Dumper operation.
* Site Dumper operation.
* 180° Excavator operation.
* 360° Excavator operation.
* Roof and wall cladding / sheeting.
* Built-up roof felting.

It is acceptable for trainees in Scaffolding (basic), Tower Crane and Excavator operation and Roof and wall cladding / sheeting to work on-site, carrying out the above work, under the direct supervision of a person who has a relevant registration card, provided they are in possession of a letter from their employer stating that they are a trainee under supervision (SHWW (Construction) (Amendment) Regulations, 2003).

2.5.3 Induction training

Source: *Proposed SHWW (Construction) Regulations, 2006, Reg.9.*

In-house induction training, provided by the employer, is mandatory for all new workers starting on a construction site.

Under section 10 of the SHWW Act, 2005, employees must receive training on commencing employment. All training and instruction must be in a form, manner and language that employees are reasonably likely to understand.

Figure 4: Topics for Induction Training

1 Company Safety Policy.
2 Company Safety Statement.
3 Site Safety & Health Plan.
4 Site team members / management.
5 Site security.
6 Emergency procedures.
7 Duties of employees.
8 Site hazards.
9 Site rules.
10 First Aid.
11 Accident reporting procedures.
12 Good housekeeping.
13 Alcohol & Drug policy.
14 PPE.

Induction training is site-specific and should give details of all topics listed in **Figure 4**, in relation to the particular site.

Records of attendance and details of topics covered in induction training should be recorded and kept on file for the duration of the project. (A sample Induction Record is available on **www.oaktreepress.com**.)

The PSCS should receive the following information at induction training:

* Name of employee.
* Name of employer.
* Contact details for employer.
* Next-of-kin contact details.
* Safe Pass Card Number.
* CSCS Card Number.

2.5.4 Toolbox Talks

Toolbox Talks are meetings, typically 15 - 20 minutes long, held on a regular basis by employers to discuss health and safety measures and concerns with employees. When used effectively, they can be the centre-pin of two-way communication and 'real' consultation. Unfortunately, in some cases, Toolbox Talks have become instructive - that is, talking at workers rather than talking to the workers and getting 'real' feed-back.

Toolbox Talks should be attended by all workers on site and should allow an opportunity for workers to raise issues regarding safety with management. The purpose is to allow input and suggestions from workers to help solve safety issues.

Topics for Toolbox Talks can be sections taken from the Safety & Health Plan or the phase of the project currently being worked on. There are no hard rules about the topics up for discussion; however, giving Toolbox Talks for the sake of it is a waste of an opportunity.

A section should be provided on Toolbox Talk sign-in sheets to record comments made, and issues raised. This will provide a record for the employer to show consultation and actions taken, should a HSA inspector require proof of consultation.

2.5.5 First Aid Training

Source: SHWW (General Application) Regulations, 1993, Part IX: First Aid; proposed SHWW (General Application) Regulations, 2006, Part VIII, Reg.41.

First Aid training is necessary for designated First Aiders in the workplace to comply with the legislation.

Typically, Occupational First Aid training lasts for three days, with an exam at the end. On successful completion; a certificate is issued to the participant. The certificate is valid for three years. However, an annual one-day refresher is recommended to keep skills up-to-date.

The HSA has approved a number of organisations to carry out Occupational First Aid training, including:

* Civil Defence School.
* Irish Red Cross Society.
* National Ambulance Training School.
* Order of Malta Ambulance Corps.
* St. John Ambulance Brigade of Ireland.
* Other individual qualified trainers on the National Ambulance Training School register.

When organising a First Aid course for employees, it is essential to ensure the competency and qualifications of the trainer / training provider. The National Ambulance Training School maintains a register of all qualified instructors / examiners. (For contact details, see **Chapter 4: Construction Health & Safety Directory**.)

2.5.6 Manual Handling Training

Source: SHWW (General Application) Regulations, 1993, Part VI: Manual Handling.

> Note: **Manual handling accounts for almost 30% of all workplace accidents.**

Manual handling training should be completed by all workers on site. Training and the practice of correct lifting techniques are key to reducing the risk of injury. Training should be specific to the tasks carried out on a regular basis by workers.
    Manual handling training courses typically contain information on:

* The law and manual handling.
* The human body.
* Guidance on fitness for the task, both mental and physical.
* Specific manual handling activities - as per the company's Safety Statement.
* How to recognise hazardous loads.
* Ways and means of avoiding or reducing the need for manual handling.
* Good handling techniques and practice in these techniques.
* Procedures for dealing with unfamiliar loads.
* The PPE to be worn.

Manual handling training can be carried out in-house, if a qualified person is available, or provided externally. When organising a manual handling course for employees, it is essential to ensure the competency and qualifications of the trainer / training provider.

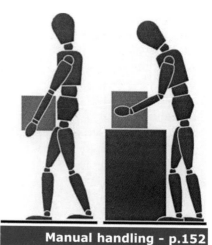

Manual handling - p.152

2.5.7 Safety Representative Training

Source: SHWW Act, 2005, s.25; *proposed SHWW (Construction) Regulations, 2006, Reg.7.*

A Safety Representative must be appointed where there are 20 or more people working on a construction site at any one time or at any stage of the project (see **2.6.4**).

A prerequisite of being able to carry out the functions of a Safety Representative is that the person is trained. Grants are available from FÁS for the training of Safety Representatives by a recognised body, such as CIF, IOSH and many other training organisations. Typically, courses run for three days and include:

* Understanding the need to work together in Health & Safety.
* Safety and the Law (Acts, Regulations & Codes of Practice).
* The role and function of the Safety Representative in inspections, accident prevention and investigations.
* Hazard identification and risk assessment.
* Basic communication skills and dealing with people.
* Recording accidents and dangerous occurrences.
* Common hazards of the construction industry.
* How to deal with safety issues on site.
* Sourcing safety and health information.

This **Handbook** provides a useful ongoing reference for Safety Representatives.

2.5.8 Training Records

Records for all training carried out in relation to Health & Safety should be kept and maintained on site at all times. Copies of certificates should be made and the originals returned to the worker. All records are required to be kept for 10 years and must be available for inspection by a HSA Inspector, if required.

2.5.9 Training providers / courses

A directory of Health & Safety training providers / courses and consultants is published annually by *HSR* magazine (see **Chapter 4: Construction Health & Safety Directory**).

Always check the qualifications and competency of trainers and ensure they are FÁS approved and registered.

2.6:   Managing Safety

2.6.1 The 4Cs

The PSCS is responsible for ensuring that there is an integrated approach to safety and health on site and that all sub-contractors / self-employed persons involved with the project work together to ensure the safety of all workers on the site. This involves co-ordinating the activities of all contractors throughout the duration of the construction period. Vigilance and monitoring are vital for the PSCS to be effective in managing health and safety on their projects. The PSCS must always remain committed to the management of health and safety on site, and should show a consistent interest in it.

Co-ordination
Co-operation
Communication
Consultation

Ensuring this co-ordination and co-operation between sub-contractors is an on-going task throughout the project, which should be addressed and reviewed at site meetings and should include:

* Emergency arrangements and procedures.
* Arrangements for the provision and use of plant and equipment that will be used by a number of sub-contractors (for example, cranes, hoists and scaffolding).
* Co-ordinating the work of sub-contractors, so as to minimise the effect of one activity on another from the point of view of safety and health.
* Ensuring sub-contractors receive relevant safety and health information relating to the project.

2.6.2 Site Meetings

Safety should be high on the agenda at site / project meetings and not be passed over at the end of the meeting due to lack of time.

The following matters could be included on agendas for project / site meetings:

* Review of current safety and health measures on site.
* Serious safety concerns.
* Reports of accidents, dangerous occurrences, near misses and complaints including analysis and follow-up action.
* Recommendations of any safety committee or site Safety Representatives regarding site safety.
* How effectively the Safety & Health Plan has been implemented to date.
* Reviewing and updating the Safety & Health Plan to take account of project progress and work to be undertaken before the next meeting.
* Identification of poorly-performing contractors, and actions to be taken to improve safety and health performance.

2.6.3 Consultation

Source: SHWW Act, 2005, s.26; SHWW (General Application) Regulations, 1993, Part II, Reg.12.

Consultation with employees

Safety consultation is the process whereby the employer and the employee secure co-operation on issues relating to safety, health and welfare in the workplace. The SHWW (General Application) Regulations, 1993, Reg.12, sets out a range of matters on which employees should be consulted, including:

* Any measures that may substantially affect safety and health.
* The assignment of employees to carry out protective and preventive services or emergency duties.
* The contents of the Safety Statement and Risk Assessments carried out and any revisions to these.
* Information on notifiable accidents and dangerous occurrences.
* The engagement of external Health & Safety consultants.
* The planning and organisation of training relating to safety and health for employees.
* The planning and introduction of new technologies as they affect particularly the choice of equipment, the working conditions and the working environment for the safety and health of employees.
* Changes in design, timescale and other key parameters that affect health and safety.

Consultation arrangements, depending on the size and nature of the project site, can range from formal meetings of a safety committee with an agreed agenda, to informal discussions with the immediate supervisor. Consultation must be balanced for both employees and management and should allow the views of employees working for different contractors to be co-ordinated and taken into account.

Consultation with other contractors

The PSCS must specify in the Safety & Health Plan the arrangements for consultations between the different contractors. Contractors are obliged to co-operate with the PSCS in ensuring there is successful consultation between the various contractors and groups of employees on the site.

It is important that, no matter what structure is implemented, consultation arrangements should include balanced participation on the part of both employees and employers, including:

* A commitment from site management to provide the necessary resources.
* Site management and employees should be encouraged to participate.
* Employees should be encouraged to communicate their views or complaints.

* Sensible recommendations should be implemented without delay.
* Site management should not ignore reasonable recommendations.
* Safety Representatives should be adequately trained and informed on safety and health matters.
* Meetings should be held regularly, in accordance with agreed procedures.
* The agenda for meetings should be varied and relevant.
* Site safety meeting attendees should be prepared to consider new options or approaches to problems.

Joint Health & Safety Agreements (SHWW Act, 2005, s.24)

A new provision in the SHWW Act, 2005 allows for employers and trade unions representing employees to enter into 'joint health and safety agreements'.

Where trade unions reach such an agreement with an employer, they may apply to the HSA to approve it. If the HSA approves such an agreement, inspectors will take account of it when assessing an employer's compliance with the legislation.

2.6.4 Safety Representatives

Source: SHWW Act, 2005, s.25; SHWW (Construction) Regulations, 2001, Reg.7.

Safety Representatives are employees elected by other site employees to consult with and make representations on their behalf to the employer / main contractor on matters relating to safety, health and welfare.

The intention of these consultations / representations is to:

* Prevent accidents and ill-health.
* Help to highlight problems.
* Help to identify solutions.

The Safety Representative has a right to:

* **Information:** The site Safety Representative has a right to access information from the PSCS regarding any health, safety and welfare issues on the construction site (excluding information of a personal or medical nature).

* **Make representations:** If he / she is of the view that there is serious or imminent danger to a person or persons, the site Safety Representative must report it immediately to the relevant management party.
* **Liaise with HSA:** A Safety Representative can communicate with, and receive information or advice from, HSA at any time. A Site Safety Representative must be advised by the PSCS if a HSA Inspector visits the site.
* Carry out inspections and investigations.
* Investigate accidents and dangerous occurrences. However, they must not interfere or obstruct the fulfilment of a statutory

obligation - for example, the scene of a fatal accident must be preserved until HSA has carried out an investigation.

* Receive adequate training required to carry out their role (see **2.5.7**).
* Reasonable time off work without loss of earnings, to acquire training that will enable them to function effectively.
* Not suffer any disadvantage or suffer any loss of earnings through carrying out their role.

> Note: **A site Safety Representative will incur no criminal liability arising from his / her performance of, or his / her failure to perform, any functions of a Safety Representative.**

Election / Selection of Safety Representatives

The PSCS has a duty to co-ordinate consultation between the contractors and their respective employees and to facilitate the appointment of a site Safety Representative, when more than 20 workers are normally employed at any one time on the site, at any stage of the project.

It is important that a site Safety Representative should represent all workers on a site, irrespective of who their direct employers are. Workers can elect the site Safety Representative at any time after site start-up. All the workers on a site, irrespective of their direct employer, are entitled to vote.

The steps involved in the election of a site Safety Representative are:

Step 1 The PSCS must facilitate the advertising of the role and function of the site Safety Representative. The employees or persons on site must elect a Safety Representative, where possible. Any worker on the site may volunteer. However, a site Safety Representative must be selected for each project by the workers on that site. A site Safety Representative moving from a site under completion to a new project is not automatically the site Safety Representative for the new site. Should more than one person volunteer, the PSCS is required to facilitate an election process by organising nominations and conducting a ballot.

Step 2 If, after advertising the role and function of the site Safety Representative, no workers volunteer for the role, the PSCS may provisionally appoint a site Safety Representative. Should 50% of the workers on the site organise an election at a later date, the outcome of that election decides the site Safety Representative and the provisional appointment is ended.

Step 3 The PSCS will provide all persons who are at work on the site at the time of the selection, and subsequently, with the name of the site Safety Representative. The PSCS must keep records of the name(s) of the site Safety Representative elected and the election / selection process carried out and make them available for inspection by an HSA Inspector.

2.7   References

* BOMEL Ltd. (2004). *Improving Health & Safety in Construction. Phase 2: Depth & Breadth. Volume 1.* Summary Report, Research Report 231 / 2004. London: HSE.

* Brabazon, P., Tipping, A., Jones, J. (2000). *Construction Health & Safety for the New Millennium*, Contract Research Report 313 / 2000. London: HSE.

* Brett, McG. (2001). *Consultation in Construction.* Safety Representatives Facilitation Project (in association with CIF / ICTU / ETS). Dublin: Joint Safety Council for the Construction Industry.

* Byrne, R. (2001). *Safety, Health & Welfare at Work in Ireland: A Guide.* Cork: Nifast.

* Garavan, T.N. (2001). *The Irish Health & Safety Handbook*, 2nd edition. Cork: Oak Tree Press.

* HSA (1994). *Guidelines on Safety Consultation & Representatives.* Dublin: HSA.

* HSA (1995). *Guide to the SHWW Act, 1989 & the SHWW (General Application) Regulations, 1993.* Dublin: HSA.

* HSA (1995). *Guidelines to the Safety, Health & Welfare (Construction) Regulations, 1995.* Dublin: HSA.

* HSA (1996). *Safe to Work.* Dublin: HSA.

* HSA (1997). *The HSA & You.* Dublin: HSA.

* HSA (1998). *Guidelines on Safety Statements.* Dublin: HSA.

* HSA (1999). *Guidelines on Preparing Your Safety Statements & Carrying Out Risk Assessments.* Dublin: HSA.

* HSA (2000). *Guidelines for Clients Involved in Construction Projects: Duties under the SHWW (Construction) Regulations, 1995 & Client Good Practice.* Dublin: HSA.

* HSA (2001). *Guide to the SHWW (Construction) Regulations, 2001.* Dublin: HSA.

* HSA (2004). *Improving Safety Behaviour at Work.* Dublin: HSA.

* HSA (2005). *A Guide to the SHWW Act, 2005.* Dublin: HSA.

* HSA (2005). *A Short Guide to the SHWW Act, 2005.* Dublin: HSA.

* HSA (2005). *Health & Safety at Work in Ireland, 1992/2002.* Dublin: HSA.

* HSA (2005). *Summary of Total Injury & Illness Statistics 2003/2004.* Dublin: HSA.

* HSA (2005). *The Safe System of Work Plan.* Dublin: HSA.

* HSA (undated). *Build in Safety - A Short Guide to Good Practice & Legislation.* Dublin: HSA.

* HSA (undated). *The Absolutely Essential Health & Safety Toolkit.* Dublin: HSA.

* HSE (1995). *Designing for Health & Safety in Construction.* London: HSE.

* HSE (1995). *The Health & Safety File*, CIS44. London: HSE.

* HSE (1995). *The Role of the Client.* CIS39. London: HSE.

* HSE (1995). *The Role of the Designer*, CIS41. London: HSE.

* HSE (2000). *The Role of the Planning Supervisor*, CIS40. London: HSE.

* HSE (2001). *Health & Safety in Construction*, HSG150(rev). London: HSE.

* HSE (2002). *Permit to Work Systems*, INDG98. London: HSE.

* McMahon, M. (2002). *What is the Law? Construction Health & Safety.* Dublin: RoundHall Ltd.

3: CONSTRUCTION HAZARDS A TO Z

In order to make this section of **The Construction Safety Handbook** as user-friendly as possible, each of the pages that follow is structured similarly:

★ **What is ...?**
★ **Hazards → Risks**
★ **Controls → Managing the Risks**
★ **Applicable Legislation / Standards / Codes of Practice**
★ **Required Documentation**
★ **Training / Certification**
★ **Further Information.**

Not all of the pages have all of these headings but, where they do, you need to know that:

★ **Hazards → Risks** - this is not an exhaustive list of the hazards and risks that may be present on a construction site. It is merely indicative of the nature of the hazards and risks commonly found. A site safety inspection (see **Chapter 2, section 2.4.2**) will identify the specific hazards and risks relevant to your site.

★ **Controls → Managing the Risks** - these are suggested control measures and do not provide an exhaustive list. In some situations, only one or two of the control measures suggested may be necessary or appropriate while, and in other situations, additional control measures will be required. All controls should be considered in the context of their relevance to your site. Draw from your own previous experience, use a practical, commonsense approach and seek professional advice, where necessary.

ABRASIVE WHEELS

See also: Hand & Power Tools / Steel Fixing

What are abrasive wheels?

Abrasive wheels include angle-grinders and consaws.

Hazards → Risks

Shattering
Entanglement
Flying fragments / molten sparks
Losing control of the equipment   → **Fire**
Explosions
Burns
Lacerations / amputation
Eye injuries and blindness

Abrasive wheels / discs and an angle-grinder

Controls → Managing the Risks

* Operate all tools as per manufacturer's instructions.
* Only use wheels that are clearly marked as per BS 4481 Part 1 (including the minimum rpm which will not be exceeded).
* Ensure that all guards, etc. are in position and serviceable before use.
* Ensure that guards are sufficient to contain fragments of bursting wheels.
* Ensure that the correct wheel is selected for the task - do not use a cutting wheel for grinding, etc.
* Allow only qualified / competent workers to fit grinding stones / cutting discs.
* Never force a grinding stone / cutting disc to fit.
* Allow only trained employees (over 18) to operate these appliances.
* Always use 110V portable electrical grinders, with the correct transformer.
* Run all wheels without load for at least one minute after mounting.
* Always use grinders at the correct angle.
* Always wear the required PPE for eyes / ears.
* Always take the necessary fire precautions to prevent fire / explosions.
* Do not cut in the vicinity of flammable liquids.
* Never use defective tools and always report defects to foreman / Safety Officer immediately.
* Keep trailing cables off the ground and away from water.
* Do not cut in storage areas.
* Avoid cutting overhead, if at all possible.

* When cutting concrete / pavement slabs, use water to safeguard against dust, where possible.

Consaws:

* Allow only responsible / competent persons to use consaws.
* Wear eye / ear protection at all times when using consaws.
* Inspect consaws for defects and damage before use.
* Report all defects / faults / damage to site management immediately and take the consaw out of use.
* Ensure that guards are in place at all times. Do not remove / bypass safety devices.

Applicable Legislation

* Safety in Industry (Abrasive Wheels) Regulations, 1982.
* SHWW (General Application) Regulations 1993, Part IV: Work Equipment Regulations.
* SHWW (Construction) Regulations, 2001.
* SHWW Act, 2005.

Training / Certification

Abrasive Wheel training courses are available from many Health & Safety training providers - for example, as a one-day course for fitters and operators who are expected to mount and maintain abrasive wheels. Typical course contents include the relevant legislation, wheel selection, wheel storage, wheel types and markings.

Further Information

* HSE: *Safety in the Use of Abrasive Wheels*.
* HSE: *Training Advice on the Mounting of Abrasive Wheels*.

ACCIDENTS

See also: Chemicals / First Aid

What is an accident?

An accident is defined as 'an unplanned event'.
Accidents rarely have a single cause; they are a series of events and failures.
Accidents consist of many contributing factors - unsafe acts, unsafe conditions,
unsafe persons, and unsafe management.

Hazards → Risks

Incorrect treatment
Untrained staff
Lack of First Aid supplies
Body fluids   →   **Infection**
Injury
Further injuries or fatalities
Prosecution

> **If a fatal accident occurs, inform HSA immediately - by phone (01 614 7000) or online at www.hsa.ie.**

Controls → Managing the Risks

✱ Provide appropriate numbers of trained
Occupational First Aiders on each construction
site, as required.

✱ Ensure that certification of First Aiders is by a
recognised Occupational First Aid Instructor
(see **Chapter 2, 2.5.5**).

✱ Provide and maintain adequate / appropriate
First Aid equipment.

✱ Report all accidents to Site Manager / Safety
Officer immediately.

✱ Complete the on-site Accident Report Book for
each accident.

✱ Investigate all accidents / dangerous
occurrences with a view to preventing
reoccurrence.

> **Do not disturb the site after a fatal accident, unless:**
> > **Three clear days have elapsed after notification to HSA.**
> > **It has been inspected by an HSA inspector.**
> > **It is necessary to treat the casualty and / or to secure the safety of other persons.**

Reportable accidents
The SHWW General Application Regulations require an employer to notify the HSA
when an injured worker 'is prevented from performing his normal work for more
than three consecutive days, excluding the day of the accident but including any
days which would not have been working days'. This means that you count Saturday
and Sunday, even if they are days off. The reporting requirement then takes effect
on day 4 - for example, if an accident happens on Friday, Saturday is day 1, Sunday

is day 2, Monday is day 3, and, if the worker is not back to his normal work on Tuesday, this is day 4, and the accident must be reported.

Accident reporting

Accident Type	To Whom?	When?
Fatal accident	HSA Gardaí	Immediately, by telephone, followed by appropriate form (IR1) (for HSA).
Critical / potentially fatal accident	HSA	Immediately, by telephone, followed by appropriate form.
Non-fatal accidents	HSA	As soon as practicable, in most cases within two weeks of occurrence.
Any dangerous occurrence as per SHWW (General Application) Regulations, 1993	HSA	Immediately, by telephone, followed by form IR3.
Any accident that results in an employee being unable to work for more than 3 consecutive days	HSA	Immediately, by telephone, followed by form IR1.
All minor accidents / injuries	Safety Officer / Site management	ASAP

Applicable Legislation

* SHWW (General Application) Regulations 1993, Part X: Notification of Accidents & Dangerous Occurrences, Regs.59 & 60.
* SHWW (Construction) Regulations, 2001.
* SHWW Act, 2005.
* *Proposed SHWW (Construction) Regulations, 2006.*
* *Proposed SHWW (General Application) Regulations, 2006, Part VIII: First Aid; Part IX: Notification of Occupational Accidents & Dangerous Occurrences.*

> **Accident investigation includes:**
> * **Witness interviews.**
> * **Measurements & sketches / photos of scene.**
> * **Testing of materials.**
> * **Follow-up report / controls.**

Further Information

* HSA: *A Strategy for the Prevention of Workplace Accidents, Injuries & Illnesses (2004-2009).*
* HSA: *Guidelines to The SHWW Act, 1989 & The SHWW (General Application) Regulations, 1993: Part X: Notification of Accidents & Dangerous Occurrences.*
* HSA: *Guidelines to The SHWW Act, 2005.*
* HSA: *Improving Safety Behaviour at Work.*
* Oak Tree Press: *Civil Liability for Industrial Accidents*, John PM White.

ACETYLENE (WELDING GAS)

See also: Welding

Hazards → Risks

Acetylene (oxy-acetylene) is a high-pressure gas that:
* Can cause rapid suffocation.
* May form explosive mixtures in air.
* Is colourless.
* Has a low odour, and thus poor warning properties at low concentrations (it gives off a garlic-like smell).

Controls → Managing the Risks

* Adhere strictly to the suppliers' safety guidelines.
* Make available the MSDS to all users of acetylene.
* Allow only experienced and properly instructed persons to handle acetylene and to take part in welding activities.
* Never attempt to transfer acetylene from one cylinder to another.
* Always use flashback arrestors with acetylene to prevent flames travelling back into the cylinder.
* Always use acetylene cylinders in the upright position - if a cylinder has been transported horizontally, leave it upright for 12 hours before use.
* Always protect cylinders from physical damage by securing in an upright position. Do not drag, roll or drop.
* Avoid inhalation of gas - wear suitable PPE at all times.
* Never use acetylene in excess of 15 psig pressure.
* Do not store acetylene in temperatures exceeding 50° C (122° F).
* Do not remove or deface labels provided by the supplier for identification of the cylinder contents.
* Always use an adjustable strap-wrench to remove over-tight or rusted caps.
* Before connecting the cylinder, ensure that back-feed from the system into the container is prevented.
* Never insert a wrench / screwdriver / pry-bar, etc into valve-cap openings.
* No smoking while handling acetylene or cylinders.
* Keep cylinders away from combustible materials.
* Ensure adequate ventilation, especially in confined areas.
* Return all empty / unused cylinders to the supplier.

When handling cylinders:

* Wear appropriate PPE. Do not wear loose clothing when handling cylinders.
* Ensure that vertical cylinders are secured or under direct control of a worker.
* Never turn your back on a free-standing cylinder.
* Avoid uneven, sloping, slippery or vibrating surfaces when standing or churning cylinders.
* Ensure that cylinders in pallets are stable before releasing strap or chain.
* Lift cylinders using suitable hoists with certified slings and hooks.
* Exercise caution when manual handling cylinders. **Note:** Acetylene cylinders are heavier than other cylinders because they are packed with a porous material and acetone.
* Do not attempt to catch a falling cylinder - get out of the way.

Ensure cylinders are securely chained when not in use

Applicable Legislation

* SHWW (General Application) Regulations, 1993.
* SHWW (Construction) Regulations, 2001.
* SHWW Act, 2005.
* *Proposed SHWW (Construction) Regulations, 2006.*
* *Proposed SHWW (General Application) Regulations, 2006.*

Training / Certification

BOC Gas Safety Awareness Workshops. Training includes:

* Relevant legislation.
* Gas properties and hazards.
* Safe use, handling, storage and transportation.
* Best practice.

Further Information

* BOC Gases: *Safe under Pressure / Safe under Power* safety videos.
* Product Stewardship website: www.airproducts.com/productstewardship/

ALCOHOL & DRUGS

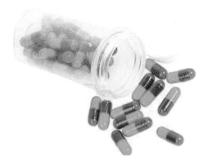

Hazards → Risks

Alcohol
Drugs
Lack of concentration
Poor performance
Lateness and absenteeism
Safety concerns
Bad behaviour or poor discipline
Accidents
Injury / death

→ **Effect on team morale and employee relations**

Controls → Managing the Risks

✶ Carry out random drug and alcohol testing on employees, if required.

✶ Ensure that employees do not consume alcohol while on duty (not even during 'business lunches').

✶ Ensure that employees do not come to work after drinking alcohol, or while still under the influence of alcohol.

✶ Ensure that employees are aware that prescription / over-the-counter drugs may influence their ability to work safely.

✶ Ensure that employees know that the use and possession of illegal substances is strictly forbidden and will result in instant dismissal (see next page for suggested contents of a workplace Alcohol & Drugs policy).

Note: **Provide employees with a drink problem with the same rights to confidentiality and support as if they had any other medical or psychological condition.**

Applicable Legislation

✶ SHWW (General Application) Regulations, 1993.
✶ SHWW Act, 2005.

Draft regulations and guidance in relation to testing are currently under review.

Further Information

✶ HSE website, Alcohol & Drugs page: http: / / www.hse.gov.uk / alcoholdrugs / index.htm.
✶ HSE: *Don't Mix It - A Guide for Employers on Alcohol.*

> **If you knowingly allow an employee under the influence of alcohol to continue working, and this places the employee or others at risk, you could be prosecuted.**

* HSE: *Drug Misuse at Work: A Guide for Employers.*
* HSE: *The Scale and Impact of Illegal Drug Use by Workers*, RR193.

Alcohol & Drugs Policy

A model workplace **Alcohol & Drug policy** would cover the following areas:

* **Aims:** Best practice is for the policy to apply equally to all grades of staff and types of work, including senior management.
* **Responsibility:** Who is responsible for implementing the policy? (**Note:** All managers and supervisors will be responsible in some way, but the policy will be more effective if a specific senior employee is named as having overall responsibility.)
* **Rules:** How does the company expect employees to behave to ensure that their alcohol consumption does not have a detrimental effect on their work?
* **Special circumstances:** Do the rules apply in all situations or are there exceptions? If so, what are these exceptions and when do they apply?
* **Confidentiality:** A statement assuring employees that any alcohol problem will be treated in strict confidence.
* **Help:** A description of the support available to employees who have problems because of their drinking.
* **Information:** A commitment to providing employees with general information about the effects of drinking alcohol on health and safety.
* **Disciplinary action:** The circumstances in which disciplinary action will be taken. What are the procedures to be followed?

ASBESTOS

What is it?

Asbestos is a common name for a group of minerals, whose fibres are very heat-resistant and strong. It was used widely until the early 1970s, when it was virtually banned in the European Union. However, there is still a lot of asbestos in buildings. Common uses of asbestos in building included:

* Filler for cement or plastic.
* Heat-resistant insulator in boilers and on pipes.
* Cladding for buildings and roofs, roof tiles, sewage and drainage pipes.
* Spray insulator on ceilings and steel girders.
* PPE.

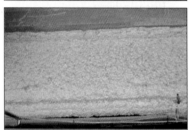

Hazards → Risks

Asbestos is hazardous to a person's health, since exposure to asbestos dust can cause lung cancer / mesothelioma. There is no known safe exposure level to asbestos - the more you are exposed, the greater the risk of developing an asbestos-related disease.

Exposure to, or inhalation of, asbestos fibres can result in cancer.

Controls → Managing the Risks

* Carry out a Risk Assessment to identify any areas of asbestos, the type and the risk of exposure.
* Provide all employees with training / information in the dangers of asbestos and the use of PPE.
* Notify HSA 28 days in advance of any work involving exposure to asbestos.
* Maintain an Occupational Health Register of all workers exposed to asbestos (**employer's duty**). Keep records for 40 years after the last entry relating to each employee (Asbestos Regs., Fifth Schedule).
* Never strip out asbestos insulation yourself. The law allows only trained and authorised personnel from specialist contractors to remove asbestos from a building.

Top to bottom: Asbestos ore; asbestos fibres; asbestos used as insulation on a girder; a worker in an asbestos suit

* If asbestos is in good condition, leave it where it is - it is only a hazard when damaged or during attempts at removal
* Keep asbestos material (including wastes) damp while working on them.
* Do not use power tools on asbestos materials, as they create dust - use hand tools instead.
* Put asbestos waste in a suitable sealed container, such as a heavy duty polythene bag, then put that bag in a second bag and label it to show that it contains asbestos.
* Ensure that disposal of asbestos waste is carried out as per the Regulations.

Applicable Legislation

* Asbestos Regulations, 1989 to 2000.
* European Communities (Protection of Workers) (Exposure to Asbestos) Regulations, 1989.
* SHWW (General Application) Regulations, 1993.
* SHWW (Carcinogens) Regulations, 2001.
* SHWW (Chemical Agents) Regulations, 2001.
* SHWW (Construction) Regulations, 2001.
* SHWW Act, 2005.
* *Proposed SHWW (Construction) Regulations, 2006.*
* *Proposed SHWW (General Application) Regulations, 2006.*

Required Documentation

Notification forms for HSA are available from http://www.hsa.ie/files/ file_20040610115318asbestosnotification.pdf.

Further Information

* HSA: *Safety with Asbestos*, 1997.
* HSE: *Selection of Suitable RPE for work with Asbestos*, HSG53.
* HSE: *Working with Asbestos in Buildings*, INDG289.
* HSE: *Asbestos Dust Kills - Keep Your Mask On: Guidance for employees on wearing RPE for work with asbestos*, INDG255 (rev1).
* Asbestos Removal Contractors Association (ARCA) - an association of specialist contractors committed to the safe removal of asbestos and other hazardous materials.
* Environmental Protection Agency - EPA deals with the licensing of storage for hazardous waste, like asbestos. If you have asbestos waste that you need to dispose of, contact your local authority immediately.
* Office of Public Works - OPW is responsible for monitoring asbestos in all government buildings.

ASTHMA

See also: Dust

What is asthma?

Occupational asthma is caused by breathing in dusts or chemicals called 'respiratory sensitisers', which are found in many workplaces.

Sensitisers common to construction include:

* Isocyanates - used in spray painting.
* Wood dusts.
* Some glues / resins.

Hazards → Risks

Irreversible allergic reaction, called 'sensitisation'. Symptoms include runny / stuffy nose, watery / prickly eyes, coughing, wheezing and chest tightness

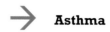 **Asthma**

Controls → Managing the Risks

* Ensure that a complete Risk Assessment and Hygiene Audit is carried out by a competent person.
* Monitor exposure levels - concentration, time periods.
* Adhere to OELs (see *Code of Practice*) at all times.
* If possible, use engineering controls to segregate substances from workers and / or use local exhaust ventilation methods.
* Where workers are exposed, provide medical examinations to monitor health.
* Provide suitable PPE to all workers.
* Ensure that MSDSs are available for all chemicals, especially 'respiratory sensitisers' (*per* Reg.42, those that 'may cause sensitisation by inhalation').

Applicable Legislation / Code of Practice

* SHWW (General Application) Regulations, 1993.
* SHWW (Carcinogens) Regulations, 2001.
* SHWW (Chemical Agents) Regulations, 2001.
* SHWW (Construction) Regulations, 2001.
* SHWW Act, 2005.
* Code of Practice for the SHWW (Chemical Agents) Regulations, 2001.
* *Proposed SHWW (Construction) Regulations, 2006.*

* *Proposed SHWW (General Application) Regulations, 2006.*

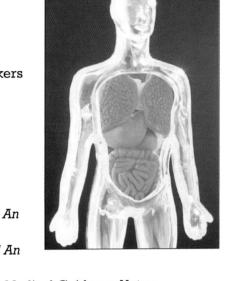

Training / Certification

* Provide training in the use of PPE to all workers exposed to, or working with, high-risk substances.
* Provide information on the risks to health / symptoms / reporting procedures.

Further Information

* HSA: *Occupational Asthma. Save your Breath! An Employees' Guide.*
* HSA: *Occupational Asthma. Save your Breath! An Employers' Guide.*
* HSA: *Occupational Asthma. Save your Breath! Medical Guidance Notes.*

* www.asthmasociety.ie
* www.asthmacare.ie

NOTES

AUGERS / DRILLS

See also: Hand & Power Tools

Hazards → Risks

Digging holes, resulting in
fractures of pipes or cables,   →   **Explosion**
or flying objects   **Electrocution**
                                       Injury

Controls → Managing the Risks

* Allow only trained operatives to operate
  augers / drills.
* Before operating augers / drills, read,
  understand and follow the
  manufacturer's operating manual and
  safety decals on the equipment.
* Check that the correct shear bolt (for
  hardness and length), as specified by
  the equipment manufacturer, is being
  used.
* Ensure that the auger / drill point and
  cutting edges are in good condition and
  the equipment is in good working order.
* Position oneself not to be hit by the
  handle, if the auger / drill stops
  abruptly.
* Operate the auger / drill at slow speeds.
* Dig the hole in several steps by clearing
  the soil frequently. Removing the soil
  reduces the load on the digger and allows for
  better control.

**An auger (above) and
an electric power drill
(next page)**

* Shut off the drive and stop the power source if the auger / drill jams.
* Watch out for loose shirts, coats and bootlaces that may get caught.
* Turn off the power before cleaning out twigs or grass.
* Do not lock the drive control in an 'on' position.
* Ensure that no one is in contact with, or near, the auger / drill before
  operating.
* Ensure that all machine guards and shields are in place before digging.

Applicable Legislation

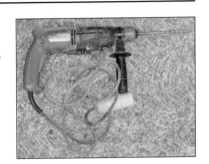

* SHWW (General Application) Regulations 1993, Part IV: Work Equipment Regulations.
* SHWW (Construction) Regulations, 2001.
* *SHWW Act, 2005.*
* *Proposed SHWW (Construction) Regulations, 2006.*
* *Proposed SHWW (General Application) Regulations, 2006.*

Further Information

* Manufacturers' / suppliers' operators manual and guidelines.

NOTES

BATTERIES

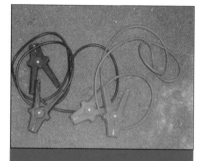

Hazards → Risks

Fire / explosion   →   **Burns**

Toxic fumes   →   **Asphyxia /
unconsciousness /
death**

In addition, the charging of batteries gives off flammable hydrogen gas. If this is allowed to collect and a source of ignition is present, it will cause an explosion, leading to acid burns.

Controls → Managing the Risks

* Adhere strictly to manufacturers' instructions for charging batteries.
* Allow only competent / qualified workers to supervise operations involving batteries.
* Ensure that recharging takes place in a well-ventilated area, with suitable fire extinguishers available.
* Switch off the battery charger before connecting / disconnecting the clips.
* Ensure that charging leads are firmly and securely clamped in position before switching on the charger.
* Adjust vent plugs before charging, as per manufacturer's instructions.
* Do not exceed recommended rate of charging.
* Turn off ignition switches before connecting / disconnecting. Always disconnect the earth first and reconnect it last using insulated tools.
* Wear suitable PPE (goggles / gloves / overalls).
* Do not wear metallic items on hands / wrists / neck.
* Do not rest tools or metallic objects on the top of the battery.
* Restrict access to the battery charging area.
* No smoking or naked lights in the charging area.

When jump-starting a battery, always:

> Ensure that both batteries have the same voltage rating.
> Check the earth polarity on both vehicles.
> Ensure the vehicles are not touching.
> Turn off the ignition on both vehicles.
> Use purpose-made, colour-coded jump leads with insulated handles - RED for the positive cable, BLACK for the negative cable.
> Ensure the leads are well clear of moving parts and that the exposed metal parts do not touch each other or the vehicle body.

Applicable Legislation

* SHWW (General Application) Regulations, 1993, Part IV: Work Equipment Regulations.
* SHWW (Construction) Regulations, 2001.
* SHWW Act, 2005, Part III: Use of Work Equipment.
* *Proposed SHWW (Construction) Regulations, 2006.*
* *Proposed SHWW (General Application) Regulations, 2006.*

Further Information

* HSE: *Electric Storage Batteries*, http://www.hse.gov.uk/pubns/indg139.htm.

NOTES

BLOCKS / BRICKS

See also: Manual Handling / Scaffolding

Hazards → Risks

Poor posture   →   **Back injury / strains and sprains**

Sharp edges / skin contact   →   **Accidents / injuries / lacerations**

Collapse of block stack /
falling block(s)   →   **Crushing**

Controls → Managing the Risks

* Stack blocks close to where they will be used.
* Stack blocks on a firm, level base and, where possible, without double-stacking of block stacks.
* Keep manual handling of blocks to a minimum and use mechanical lifting / handling aids, where possible.
* Ensure that the landing area of the working platform is adequate for the temporary loading of blocks - for example, a dedicated loading bay with a known SWL.
* Arrange work so that blocks can be handled close to the body, and with access from all sides of the block stack, where possible.
* Arrange work to avoid over-reaching or twisting when handling the blocks, and so that blocks only need be handled up to shoulder-height, where possible.
* Ensure good grip and secure footing in the work area when handling blocks.
* Keep work areas free from anything that will contribute to slips, trips and falls, especially on scaffolding.
* Exercise care when using wall ties on double-skin walls, due to sharp edges.
* Wear PPE such as safety helmets / footwear and protective gloves / overalls.
* Keep blocks covered, where possible, to avoid them absorbing rain / moisture and thus increasing their weight.

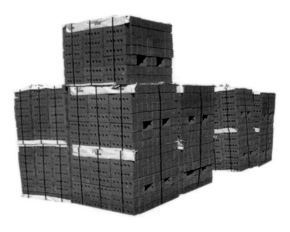

Applicable Legislation

* SHWW (General Application) Regulations, 1993, Part VI: Manual Handling of Loads, Regs.27 & 28; Eighth Schedule.
* SHWW (Construction) Regulations, 2001.
* SHWW Act, 2005.
* *Proposed SHWW (Construction) Regulations, 2006.*
* *Proposed SHWW (General Application) Regulations, 2006, Part V.*

Training / Certification

* Provide manual handling training annually by a competent trainer.
* Keep records of training on file on-site.

Further Information

* HSA: *Handle with Care - Safe Manual Handling.*
* HSE: *Handling Heavy Building Blocks*, CIS37.

DANGER
Wet blocks increase in weight by 25%. Keep blocks dry.

NOTES

BULLYING

See also: Harassment / Stress

What is bullying?

"Bullying in the workplace is repeated inappropriate behaviour, direct or indirect, whether verbal, physical or otherwise, conducted by one or more persons against another or others, at the place of work and / or in the course of employment, which could reasonably be regarded as undermining the individual's right to dignity at work" (HSA *Code of Practice on Bullying*).

Examples of behaviour that may be classified as bullying include:

* Purposely undermining someone.
* Victimisation.
* Humiliation.
* Manipulation of an individual's reputation.
* Social exclusion or isolation.
* Intimidation.
* Verbal or physical abuse or threats of abuse.
* Aggressive or obscene language.
* Jokes that are obviously offensive to one individual, by spoken word or email.
* Intrusion by pestering, spying and stalking.
* Unreasonable assignments to duties that are obviously unfavourable to one individual.
* Repeated requests with impossible deadlines or impossible tasks.

Hazards → Risks

Emotional illness
Fear
Anxiety
Depression   →   **Stress**
Bodily harm
Physical assault
Absenteeism

Controls → Managing the Risks

* Adopt a zero-tolerance Anti-Bullying policy.
* Adhere to the company's Anti-Bullying policy.
* Use the company's Grievance policy / procedures to deal with all complaints.

* Report all incidents of bullying to the Safety Officer, who will investigate immediately and take the necessary corrective action.

Applicable Legislation / Code of Practice

* SHWW (General Application) Regulations, 1993.
* S.I. No.17 of 2002. Code of Practice detailing Procedures for Addressing Bullying in the Workplace (LRC Code).
* S.I. No.72 of 2002. Code of Practice on Sexual Harassment & Harassment at Work (Equality Authority Code).
* Equality Act, 2004.
* SHWW Act, 2005.
* *Proposed SHWW (General Application) Regulations, 2006.*

* HSA: *Code of Practice on the Prevention of Workplace Bullying.*

Required Documentation

* Anti-Bullying policy (see **www.oaktreepress.com** for sample contents).
* Grievance / Complaints policy & procedures (see **www.oaktreepress.com** for sample contents).

Further Information

* HSA: *Bullying at Work.*
* HSA: *Report of the Task Force on the Prevention of Workplace Bullying.*

* Bullying Response Unit, HSA.
* Equality Authority.
* Labour Relations Commission.

CARTRIDGE-OPERATED TOOLS

See also: Hand & Power Tools

What are cartridge-operated tools?
The most-common form of cartridge-operated tool found on most building sites is a nail gun (see picture on next page).

Hazards → Risks

Accident / injuries, including
puncture wounds
Blindness

Eye injuries

Controls → Managing the Risks
* Maintain and operate tools as per manufacturer's instructions.
* Allow only competent / qualified (over 18) persons, not suffering from colour blindness, to operate tools.
* Wear the required PPE (helmet / goggles / ear protection).
* Before starting, always check suitability of material for cartridge fixing. Do not fire into unfamiliar materials without a trial fixing using a low-powered cartridge or hammer.
* Always check the area behind the material / structure into which fixing is being fired before commencing.
* Always carry out a fire test.
* Control access to the area where work is being carried out.
* Ensure that tools are at right angles to the surface.
* Ensure that the whole rim of the splinter-guard is firmly against the workface, so as to stabilise the tool and leave no gaps.
* Secure the gun and cartridges, so as to prevent unauthorised use.
* Ensure that the issue / use of tools is strictly controlled.
* Store tools unloaded in secure, cool and dry stores.

Do not:
> Use a suspect tool.
> Use force when loading cartridges.
> Load a cartridge before you need it.
> Leave a loaded tool lying about.
> Point a tool at other people.
> Fix where another fixing has failed.
> Strip a tool down when it is loaded.
> Press tool against your hand.
> Leave cartridges about job site.
> Insert cartridges before the nail or stud is loaded.
> Fool about, or wave tool in the air.

* Mark cartridges of different strengths clearly and store them separately.
* Never use a cartridge-operated tool in a combustible environment or in the presence of flammable liquids / materials.
* Do not expose tools to extreme temperatures. Only use in a well-ventilated area.

Applicable Legislation

* SHWW (General Application) Regulations, 1993, Part IV: Work Equipment Regulations.
* SHWW (Construction) Regulations, 2001.
* SHWW Act, 2005.
* *Proposed SHWW (Construction) Regulations, 2006.*
* *Proposed SHWW (General Application) Regulations, 2006, Part III: Use of Work Equipment.*

Further Information

* Manufacturers' and suppliers' guidelines.

NOTES

CEMENT MIXERS

See also: Concrete & Cement / Machinery - Moving Parts

Hazards → Risks

Entanglement
Skin contact with cement
Trapped fingers, hands or hair **Crush injuries, leading
to amputation / trauma**
Burns / lacerations

Controls → Managing the Risks

* Adhere strictly to manufacturer's instructions.
* Ensure that no modifications or make-shift repairs are carried out.
* No bypassing or interfering with safety devices on mixers.
* Allow only trained / authorised workers to service mixers.
* Exercise care when loading, to avoid the shovel getting caught on the drum.
* Ensure that the drum is stationary and that the mixer is turned-off / isolated before any attempt is made to clean it.
* Report all defects immediately to site management and shut down / isolate the mixer so as not to endanger other workers.
* Operate mixers only on firm, level ground.
* Never mount mixers on blocks.
* Wear suitable PPE - hand protection / eye protection / overalls.

Applicable Legislation

* SHWW (General Application) Regulations, 1993, Part IV: Work Equipment Regulations.
* SHWW (Construction) Regulations, 2001.
* SHWW Act, 2005.
* *Proposed SHWW (Construction) Regulations, 2006.*
* *Proposed SHWW (General Application) Regulations, 2006.*

Further information

* Manufacturers' / suppliers' instructions and guidelines.

Dumpers / Dump Trucks - see p.92

CHAINS & SLINGS

See also: Lifting Equipment & Operations / Teleporters / Working at Height

Hazards → Risks

Unsuitable / damaged chains or slings
Incorrect swinging
Untrained banksman / slinger
Loads falling   →   **Injury**
Crush injuries
Head injuries
Death

Controls → Managing the Risks

* Ensure that all lifting equipment used on site is CE-marked, or certified as designed to a relevant recognised international standard, particularly in relation to protecting workers from falling objects and overturning.
* Risk-assess all new lifting equipment prior to use, with particular reference to the proposed area of use.
* Ensure that all lifting equipment is inspected and certified by a competent person. Re-test / re-certify every six months. Keep all certificates on file on-site.
* Ensure that the supplier provides relevant documentation declaring equipment fit for work (**supplier's responsibility**).
* Label and take out of service all defective lifting equipment, until it is repaired and re-certified by a competent person.
* Replace hooks, rings and eyebolts if damaged.
* Replace worn / stretched chains or slings.
* Mark and remove off-site all damaged chains and slings.
* Ensure that the SWL is marked on all chains and slings. Do not exceed the SWL marked on chains and slings under any circumstances. Do not overload slings / chains.
* Do not wrap chains and slings around the forks of a teleporter when lifting loads.
* When using chains or slings with forks, use suitable fork clamps, with the chain or sling suspended from a suitable hook or shackle.
* Do not carry out lifting operations in bad weather, or near overhead power lines.

* Ensure that all loads are slung by a trained and competent person (banksman).

Applicable Legislation

* Factories Act, 1955, s.34 (1) (a).
* Chains, Ropes & Lifting Tackle (Register) Regulations, 1956.
* European Communities (Wire Ropes, Chains & Hooks) Regulations, 1979.
* Safety in Industry Act, 1980.
* SHWW (General Application) Regulations, 1993, Part IV: Work Equipment Regulations.
* SHWW (Construction) Regulations, 2001, Part 15: Chains, Ropes & Lifting Gear, Regs.105-113.
* SHWW (General Application) (Amendment) Regulations, 2001.
* SHWW Act, 2005.
* *Proposed SHWW (Construction) Regulations, 2006.*
* *Proposed SHWW (General Application) Regulations, 2006, Part III: Use of Work Equipment; Part XV: Work at Height.*

Required Documentation

All slings / chains must be inspected by a competent person (fitter) at least once every six months and the necessary documentation completed:

* CR 4A - Lifting Appliances - Report of thorough examination (14 months / repair / first use)
* CR 4B - Lifting Appliances - Weekly inspection report
* CR 6 - Chains, Slings, Rings, Links, Hooks, Plate clamps, Shackles, Swivels & Eyebolts - Certificate of test & examination
* CR6A - Chains, Ropes & Lifting gear - Report of thorough examination
* CR6B - Chains, Ropes & Lifting Gear - Report of annealing / heat treatment

Training / Certification

* FÁS CSCS Telescopic Handler operation.
* FÁS CSCS Mobile Crane operation.
* FÁS CSCS Tower Crane operation.
* FÁS CSCS Slinger / Signaller.

Further Information

* Manufacturers' / suppliers' guidelines and information.

CHAINSAWS

See also: Hand & Power Tools / Machinery - Moving
Parts

Hazards → Risks

Moving parts / blade
Incorrect / insufficient PPE
Inexperienced workers   →   **Serious injury - amputation /**
Lapse of concentration   **severe bleeding**
Faulty equipment / parts
Unconsciousness

Controls → Managing the Risks

* Allow only competent / qualified workers to operate chainsaws as per the operator's handbook.
* Maintain / service chainsaws as per the manufacturer's instructions.
* Wear the following PPE:
  o Safety helmet (to conform to BS 5240) complete with visor and earmuffs, or hearing protection, and eye protection.
  o Clothing - close fitting.
  o Gloves - with protective pad on the back of the left hand.
  o Leg protection - incorporating loosely woven long nylon fibres or similar protective material (kevlar).
  o Chainsaw boots with steel toecap and kevlar in the instep and the side.
* Always ensure that the blade is stationary when idling.
* Do not use the saw above shoulder-height.
* Walk with the blade behind you when carrying a chainsaw.

To prevent kickback when using chainsaws:
> Ensure that proper chain and chainsaw maintenance is carried out.
> Never begin cutting with the upper half of the nose of the blade.
> Watch out for branches, logs or other material that could come in contact with the danger zone while cutting.
> Grip the chainsaw properly using both hands, the thumb of the left hand should be under the handle.
> Keep the left arm straight before cutting - in the event of kickback, the saw is likely to be diverted over one's body.
> Do not attempt to cut material above shoulder-height.
> Do not run the engine slowly at the start or during a cut.
> Consult the chainsaw user's handbook for further information.

* Apply the chain-brake when the saw is not cutting (let engine revs drop to idle and apply the chain-brake with the back of the left hand).
* Ensure that any other person in the vicinity is at least two saw-lengths away from the operator.
* Check periodically that all nuts / bolts are secure.
* Do NOT work alone.
* Ensure that a First Aider is available while a chainsaw is in use.
* Never leave chainsaws unattended.
* Do not use chainsaws in conditions of failing daylight, especially in dusk or darkness.

Most construction contractor's insurance policies do not cover the use of chainsaws as standard.

Before use, ensure the correct insurance requirements are met.

Applicable Legislation / Standards

* SHWW (General Application) Regulations 1993, Part IV: Work Equipment Regulations.
* SHWW (Construction) Regulations 2001.
* SHWW Act, 2005.
* *Proposed SHWW (Construction) Regulations, 2006.*
* *Proposed SHWW (General Application) Regulations, 2006, Part III: Use of Work Equipment.*

* IS / EN 381:1993: Protective clothing for users of hand-held chainsaws (Parts 2, 3, and 5).
* IS / EN 381:1996: Protective clothing for users of hand-held chain saws (Part 2).

Further Information

* HSA website: Industry Sectors > Agriculture & Forestry > Major Hazards - Farms > Preventing chainsaw accidents.

* HSA: *Safety in Forestry Operations.*

CHEMICALS

See also: Acetylene / Concrete & Cement / Solvents / Welding

What are hazardous chemicals?

Hazardous chemical agents are defined as:

* Those meeting classifications for dangerous substances in Directive 67 / 548 / EC.
* Those that, because of physio-chemical, chemical or toxicological properties and / or the way in which they are used or are present in the workplace, may present a danger to the health and safety of workers.

Hazardous chemical agents that are typically present within a construction site include:

* Cement / concrete.
* Asbestos.
* Lead.
* Welding fumes.
* LPG.
* Solvents, paints and cleaning agents.

Hazards → Risks

Incorrect handling, use or storage	**Contamination**
Inhalation	**Ill-health**
Spillages	**Dermatitis**
Accidental release	**Burns**
Lack of information	**Poisoning**
	Death

Controls → Managing the Risks

* Comply with all requirements of the relevant MSDS.
* Keep all MSDS in the on-site Safety File, and a copy with the Safety Officer.
* Ensure that there is good ventilation in areas where work is being carried out.
* Observe good personal hygiene habits.
* Wear suitable PPE.

> **Procedures must be in place to deal with emergencies / accidents / incidents and the persons responsible for such procedures must be identified.**

* Inform workers of any materials that may be harmful.
* Provide training for the safe handling / use / storage and emergency procedures for all workers.

> **Remove all contaminated PPE / clothing at once and wash separately.**

* Label, store and transfer all chemicals properly, as per the relevant MSDS.
* Dispose of all contaminated materials / waste, as per the relevant MSDS.
* Highlight First Aid requirements on MSDS for all chemicals.
* Adhere to OEL values (see *Code of Practice*) at all times.
* Provide adequate washing facilities / eye-wash stations.
* Ensure that adequate numbers of qualified First Aiders are present on site.

Common on-site chemicals

Substance	Used for	Hazard	Managing the Risk
Xylene	Chemical cleaner	Dermatitis	Gloves, goggles, good ventilation, washing facilities. No smoking.
Glue / Adhesive	Roof work / wood work		Gloves, eye protection, good ventilation, face-washing. No smoking.
Cement	Masonry / plastering work in particular	Dermatitis / Skin burns / harmful lime content	Gloves, boots, personal hygiene, protective creams. Washing facilities, eye-wash stations.
Mineral Fibre (rockwool) / Fibre Glass	Insulation work	Dermatitis	Gloves, overalls, goggles, masks. Minimise cutting and handling as much as possible.
Solvents - xylene, tri-chlorethane etc	Cleaning products and products used in roof work	Dermatitis	PVC gloves at all times. Well-ventilated areas.
Acids - Hyplrochloric, hydroflouric and sulphuric	Masonry cleaning	Dermatitis	Skin and eye protection. Personal hygiene.
Site contaminants	Site re-develop-ment, involving ground work, demolition, tunnelling, excavations, etc	Present in soil, arising from previous industrial activities / microbiological risks include: Weils Disease, Tetanus, Hepatitis B	Protective clothing. Good personal hygiene. Respiratory protection. Adequate washing / First Aid facilities.
Bituthene Prepre	Water-proofing	Dermatitis	Gloves, personal hygiene.

Classification of chemicals

C	Corrosive	
E	Explosive	
F F+	Highly Flammable Extremely Flammable	
N	Dangerous for the Environment	
O	Oxidising	
T T+	Toxic Very Toxic	
XI Xn	Irritant Harmful	

Applicable Legislation / Code of Practice

* SHWW (General Application) Regulations, 1993.
* SHWW (Carcinogens) Regulations, 2001.
* SHWW (Chemical Agents) Regulations, 2001.
* SHWW (Construction) Regulations 2001, Part 9: Regs. 36-38, Dangerous or unhealthy atmospheres.
* SHWW (Explosive Atmospheres) Regulations, 2003.
* SHWW Act, 2005.
* *Proposed SHWW (Construction) Regulations, 2006.*
* *Proposed SHWW (General Application) Regulations, 2006.*

Health surveillance must be available when workers are exposed to a hazardous chemical, where there is an identifiable disease, adverse health effect or reasonable likelihood of disease associated with working with the chemical.

* HSA: *Code of Practice for the SHWW (Chemical Agents) Regulations, 2001*.

Further Information

* HSA - *Chemical Legislation: An Overview*.
* HSA - *Risk Assessment of Chemical Hazards*.
* HSE: *Chemical Cleaners, CIS24(rev.1)*.

Chemicals should always be stored securely

NOTES

COLD (HYPOTHERMIA)

See also: Welfare

What is hypothermia?

* **Primary hypothermia** - occurs when the body's heat balancing mechanisms are not working properly and are subjected to extreme cold.
* **Secondary hypothermia** - where heat-balancing mechanisms are impaired and cannot respond adequately to moderate cold.

Hazards → Risks

Cold / exposure

Wet / rain / wind   →   Hypothermia

Controls → Managing the Risks

* Provide employee training in the prevention, recognition and treatment of hypothermia.
* Provide suitable protective clothing.
* Maintain all heating devices in plant / cranes / machinery as per manufacturer's instructions.
* Provide adequate / suitable means of drying wet clothing, as well as shelter from bad weather.

First Aid measures

* Prevent further heat loss - move casualty indoors, if possible.
* Insulate the body well, using clothing, plastic and body heat.
* Remove wet clothing, only if dry replacements are available.

Do not:

* Give the casualty alcohol.
* Rub the casualty's body.
* Heat using artificial heat sources placed directly on the body.

Symptoms of hypothermia include:

> Cold, pale skin.
> Intense shivering.
> Speech becomes slurred.
> Muscles become rigid.
> Disorientation.
> Effects on the nervous system - poor co-ordination.
> Clumsiness / sleepiness / amnesia.
> Irrational behaviour.

Applicable Legislation

* SHWW (General Application) Regulations, 1993, Parts IV & V.
* SHWW Act, 2005.
* *Proposed SHWW (General Application) Regulations, 2006.*

Further Information

* Irish Water Safety - information on hypothermia.

NOTES

COMPANY VEHICLES

6 people were killed in 2004 by vehicle movements on construction sites

Hazards → Risks

Slippery and / or unsuitable surfaces
Poor visibility
Unqualified and / or inexperienced drivers
Inadequate and / or restrictive insurance policies
Unsuitable / faulty vehicles
Over-loading
Passengers

→ **Collision Injuries Death**

Controls → Managing the Risks

* Maintain all vehicles as per Regulations / Road Traffic Acts.
* Ensure that all drivers hold a current driver's licence and are covered on the company's insurance policy.
* Report any endorsements or penalty points received on licences to site management.
* Keep all site areas clear of anything that will obstruct a driver's vision and of substances that will contribute to slippery surfaces.
* Ensure that all difficult manoeuvres or reversing are controlled by CCTV or guided by a responsible person.
* Park non-construction vehicles in designated areas away from construction site traffic.
* Mark carparks clearly for contractors' / sub-contractors' / visitors' cars.
* Display speed signs throughout the site.
* Adhere to speed limits on construction sites at all times.
* Ensure that all workers comply with the site traffic plan at all times.
* Do not carry passengers in the rear of vans, unless the van is adapted suitably.
* Give right-of-way to construction plant and machines on construction sites at all times.

Separate car parks for contractors' vehicles (top); pedestrian crossings at approriate places (middle); and speed limits (above) - all contribute to on-site safety

Site traffic control

All construction sites should have a site traffic plan to control traffic movement. Ideally, this should segregate construction plant and machines from cars, and should eliminate the need for reversing, if at all possible.

The site traffic plan should be explained at Induction training and be displayed in a prominent location - for example, the Canteen.

Plans may include the use of warning signs, bollards, stop / go systems, ramps, temporary traffic lights and flags men.

For construction work close to roads or road works, liaison may be necessary with local Gardaí.

Applicable Legislation

★ Road Traffic Acts, 1961 to 2003.

Further information

★ Hibernian Insurance - Driver's Handbook for company vehicles that can be personalised to your company, http://hibernian.netsource.ie/pdfs/gen_Driver_Handbook.doc.
★ HSA: *Safety of Workplace Vehicles*.

NOTES

COMPRESSORS / COMPRESSED AIR

Hazards → Risks

Explosion	→	**Fire and burns**
Injection	→	**Injuries**
Spillage	→	**Toxic fumes**

Controls → Managing the Risks

* Maintain / operate all compressors as per the manufacturer's instructions.
* Carry out work with compressed air in accordance with the minimum requirements as specified in the Construction Regulations, 2001, Sixth Schedule.
* Allow only suitably trained workers to use this equipment.
* Ensure that all workers working in compressed air are subject to regular medical examinations for fitness and health.
* Maintain records for all workers, recording the time spent continuously working in compressed air.
* Put in place emergency procedures, specific to risk.
* Maintain all guards in position.
* Ensure that every compressor is thoroughly cleaned and examined at least once in every period of 26 months by a competent person.
* Do not use compressed air to clean up or to clean clothes.
* Adopt a zero-tolerance policy on horseplay with compressed air.
* Check all air tools / spray guns to ensure the pressure rating is compatible with the air supply line.
* Wear suitable PPE.

Applicable Legislation

* Safety in Industry Act, 1980, s.33.
* SHWW (General Application) Regulations, 1993.
* SHWW (Construction) Regulations, 2001: Part 7, (31-34); Sixth Schedule.
* SHWW Act, 2005.
* *Proposed SHWW (Construction) Regulations, 2006.*
* *Proposed SHWW (General Application) Regulations, 2006.*

Further Information

European Council Directive 84/553/EEC sets noise limits and requirements for the issue of an EC type-examination certificate for compressors.

Working at height - see p. 224

CONCRETE & CEMENT

See also: Cement Mixers / Chemicals / Formwork / Hand & Power Tools / Pre-cast Elements & Components

Hazards → Risks

Concreting - wet concrete and cement dust	→	**Dermatitis / concrete burns / occupational asthma / silicosis**
Inhalation of dust	→	**Asthma / silicosis**
Manual handling	→	**Injuries**

Controls → Managing the Risks

* Wear suitable PPE (gloves / wellingtons / overalls) during all concrete operations.
* Do not expose skin to wet concrete at any time.
* Seek immediate First Aid assistance if a worker receives concrete burns.
* Provide good welfare / hand-washing facilities.
* Provide MSDS sheets for all forms of concrete.
* Where there is a risk from silica dust during cutting operations, wear suitable RPE.
* Wear PPE at all times, including goggles to prevent eye injuries, especially with all scabbling operations.
* Provide health surveillance for workers working with wet cement on a regular basis.
* Avoid manual handling of heavy loads. Order cement only in 25kg bags. Do not allow workers to handle more than one bag at a time.
* Ensure that tamping and levelling beams are light enough to be handled safely.
* Allow only responsible / competent workers to use cement mixers.
* Operate mixers only on firm, level ground. Never mount mixers on blocks.
* Ensure that concrete is pulled / raked only from outside the wet concrete. Do not allow workers to stand / work in wet concrete, when pulling / raking.

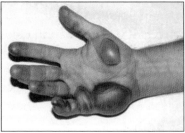

Cement burns

Warning
Water-soluble chromium VI, classified as carcinogenic and sensitising, can induce painful, disabling, allergic eczema in people exposed to wet cement preparations widely used in construction. Reducing chromium VI by adding 0.35% ferrous sulphate to the cement has been shown to reduce the adverse health effects.

When pouring concrete:

* Ensure that concrete pumping rigs are maintained and serviced on a regular basis.
* Ensure that periodic checks are carried out on outriggers, etc.
* Do not pour concrete too rapidly, or from such a height as to overload the formwork.
* Carry out periodic checks of the formwork as the pour proceeds.

When pumping concrete, exercise caution when working close to overhead lines.

Power floats / concrete vibrators

* Ensure that power floats are serviced regularly by a competent person and that records are kept of this servicing.
* Ensure that all safety guards are in place at all times on power floats / vibrators and that safety release levels are in good working order.
* When power floating close to open edges, provide edge protection.
* Ensure that the vibrator motor is on an elevated platform and is firmly secured.
* Do not allow poker vibrators to come in contact with any person.

Applicable Legislation

* SHWW (General Application) Regulations, 1993.
* SHWW (Chemical Agents) Regulations, 2001.
* SHWW (Construction) Regulations, 2001.
* SHWW Act, 2005.
* *Proposed SHWW (Construction) Regulations, 2006.*
* *Proposed SHWW (General Application) Regulations, 2006.*

**Top to bottom:
Concrete pump;
pouring concrete;
power floating concrete**

Further Information

* Irish Concrete Federation - for MSDS.

* European Agency for Safety & Health at Work: FACTS 40: *Skin Sensitisers*.
* HSE: *Cement*, CIS26 (rev.2).
* HSE: *Silica*, CIS36 (rev.1).

CONFINED SPACES

See also: Gas / Tunnelling

What are confined spaces?

A confined space:

* Is substantially enclosed.
* Has the presence, or foreseeable presence, of at least one of these hazards:
  o Flammable or explosive atmosphere.
  o Harmful gas, fumes or vapour.
  o Free-flowing solids.
  o Excess of oxygen.
  o Excessively high temperature.
  o Lack, or reasonably foreseeable lack, of oxygen.
* Offers a risk of injury arising from one or more of these hazards, where such an injury would merit emergency rescue.

Confined spaces include: Trenches, shafts, caissons, tunnels, pipelines, pits, septic tanks, sewers, manholes, vaults and bins.

Getting ready to work in a confined space

Hazards → Risks

Temperature extremes
Noise
Falling objects
Fire / explosion   →
Lack of oxygen
Flooding
Increased body temperature

Injury
Asphyxiation /
  loss of consciousness
Drowning
Death

Controls → Managing the Risks

* Before permitting entry to a confined space, carry out a complete inspection of the work area to detect harmful gases, risk of collapse, etc.
* Ensure that Risk Assessments, Method Statements and Permits to Work are all in place for work in confined spaces, and that these are reviewed and approved by PSCS / Safety Officer.
* Allow only trained / competent workers to enter confined spaces.

* Develop emergency / rescue procedures, and train all workers in these procedures before entering confined spaces.

* Ensure that a trained rescue team, and a First Aider per person, is in place at all times, during work in confined spaces.

* Use gas detectors to detect the build-up of harmful gases. Test air at all levels within the confined space - good air near the opening does not mean there is good air at the bottom!

* Use oxygen / CO2 monitors at all times in a confined space.

* Ensure that all atmospheric testing is carried out by a competent person, using a suitable gas detector which is correctly calibrated.

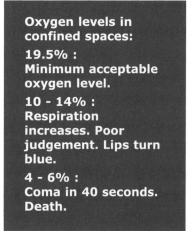

Oxygen levels in confined spaces:

19.5% : Minimum acceptable oxygen level.

10 - 14% : Respiration increases. Poor judgement. Lips turn blue.

4 - 6% : Coma in 40 seconds. Death.

* Ensure that all gas detectors are calibrated regularly.

* Put in place a safe method of access and egress, which will also allow emergency evacuation.

* Put in place a 'buddy' system when working in confined spaces. Do not allow workers to work alone close to, or in, a confined space.

* Put in place a suitable communication system at all times, to allow verbal communication between those on the surface and those in the confined area.

* Wear suitable / approved PPE at all times.

* Ensure that lifelines attached to a safety harness run back to a point / tripod outside the confined space. Ensure a rescue tripod is available at all times.

* Test all respiratory equipment / RPE before use in a confined space.

Permit to Work system
This is a system used to ensure that a safe system of work is in place. Generally such systems are used for high-risk activities such as confined space entry, demolition, etc. Permits to work allow only authorised persons to enter the work area, under very controlled laid-down conditions. All permits to work should be signed off by PSCS / Project Manager / Site Manager / Safety Officer. Permits should be valid for 12 hours only.

Tripod
A tripod (**pictured right**) is a standard piece of rescue equipment used when working in confined spaces: a worker can be lowered into the confined space by a 'buddy' and, more importantly, can be raised out of the confined space with ease.

Applicable Legislation / Code of Practice

* SHWW (General Application) Regulations, 1993.
* SHWW (Chemical Agents) Regulations, 2001.
* SHWW (Confined Space Entry) Regulations, 2001.
* SHWW (Construction) Regulations, 2001.
* SHWW Act. 2005.
* *Proposed SHWW (Construction) Regulations, 2006.*
* *Proposed SHWW (General Application) Regulations, 2006.*

* HSA: *Code of Practice for Working in Confined Spaces.*

A confined spaces gas detector

Training / Certification

FÁS Confined Space Entry training:

* C.1 - Management of risk during Confined Spaces Working - 2 days.
* C.2 - The Supervising of Risk during Confined Spaces Working - 4 days.
* C.3 - Working in Confined Spaces Safe Working Practice - 4 days.

Further Information

* HSA: *Working in Confined Spaces.*
* HSE: *Safe Work in Confined Spaces*, INDG258.

NOTES

Concrete & Cement see p. 78

CRANES (MOBILE & TOWER)

See also: **Lifting Equipment & Operations / Working at Height**

Hazards → Risks

Structural damage
Working at heights
Human error
Falling objects / materials   →
Incorrect signage
Adverse weather conditions
Electrical cables
Unauthorised access

Lacerations
Amputation, because
of entanglement
Death
Collapse of structures
Collisions

Controls → Managing the Risks

* Responsibility for safeguarding work involving cranage lies primarily with the sub-contractor supplying and erecting cranes on site.
* Do not use cranes in adverse weather conditions, such as high winds / fog / sleet etc.
* Ensure that all safety documentation is completed and inspected before work starts.
* Ensure that cranes are inspected by a competent person and the necessary documentation (see below) is completed at prescribed intervals.
* Carry out all operations as per the operator's manual.
* Ensure that the crane's engine is turned off and the hand-brake is in place before work starts.
* Maintain all guards in position. Be alert to high risk areas - rotating shafts / gearing / belts or chains slewing / counterweight.
* Allow only trained crane drivers (valid CSCS card) to drive cranes.
* Wear the required PPE.
* Do not use mobile phones while operating a crane.
* Do not allow a person under 18 to operate cranes or to give signals to the operator.
* Do not operate machinery while under the influence of alcohol / drugs, including prescribed medication.
* Do not wear loose clothing / jewellery / belts, etc.

A mobile crane (left) and tower crane (right)

* Allow only trained persons over 18 to act as banksmen and ensure a safe system is in place for all lifting operations. (Banksmen should always be clearly identifiable from other workers, especially at a height - for example, use different coloured high-visibility vests - **see picture, right**.)

* Do not carry passengers on crane.
* Do not drive mobile cranes with the jib in the raised position.
* Ensure an approved SWL indicator is used on jib cranes and that it is inspected by a competent person before use and weekly thereafter. Do not exceed SWL.
* Take heed of any warning alarm when lifting loads near SWL.
* Secure all cranes when left unattended to prevent unauthorised use, especially at the end of the working day.

Applicable Legislation / Standards / Code of Practice

* SHWW (General Application) Regulations 1993, Part IV: Work Equipment Regulations.
* SHWW (Construction) Regulations 2001, Part 11: Transport, Earthmoving & Materials Handling Machinery, Regs.41-46; Part 14: Lifting Appliances, Regs.80-99.
* SHWW Act, 2005.
* *Proposed SHWW (Construction) Regulations, 2006.*
* *Proposed SHWW (General Application) Regulations, 2006, Part III: Use of Work Equipment.*
* IS 360:2004 - *Code of Practice: Safe Use of Cranes in the Construction Industry, Part 1: General* - available from NSAI.

Training / Certification

* FÁS CSCS Tower Crane operation.
* FÁS CSCS Mobile Crane operation.
* FÁS CSCS Crawler Crane operation.
* FÁS CSCS Slinging / Signalling.

Required Documentation

* CR3 Crane - Certificate of test & examination.
* CR3A Crane - Report of anchoring / ballasting test.
* CR3B Crane - Report of automatic safe load indicator test.

DEMOLITION WORK

Hazards → Risks

Loss of control of the process -
uncontrolled or disproportionate
collapse in overloading of
a structure   →   **Fire / explosion / flooding**
Machinery operations   **Injury / Electrocution**
Debris / falling objects   **Suffocation / Death**

Controls → Managing the Risks

* Before any demolition operation commences, inspect and survey the site to ascertain the condition of structures, ground and services, including overhead / underground electricity lines and gas pipes.
* Ensure that any demolition is under the control of a competent person - for example, an engineer.
* Assess the structure for the presence of asbestos and, where asbestos is present, remove it before actual demolition work starts.
* Secure the demolition site.
* Ensure there is good communications between the demolition machine operators and other site users.
* Wear suitable PPE, including approved safety helmets.
* Take appropriate precautions to safeguard against fire, explosion, flooding, electric cables, gas pipes and noise.
* Ensure that the demolition will not affect the stability of neighbouring structures.
* Use shoring as required, to support adjacent structures and ensure it is checked regularly by a competent person as demolition proceeds.
* Suppress dust during the demolition process.
* If the demolition is being carried out on a phased basis, ensure that all storage areas are capable of bearing the additional weight.

Exclusion zones:
* Establish an exclusion zone at a distance from, and surrounding, any structure being demolished - especially if using explosives.
* Sole responsibility for designing the exclusion zone lies with the explosives engineer.

* Ensure that people outside the exclusion zone are safe from all demolition work and that they remain outside the exclusion zone.
* Ensure that, if the shot-firer needs to remain within the zone, he / she is in a safe position.
* An exclusion zone is built up from four areas:
  o Plan area - the structure to be demolished.
  o Designed drop area - the area where the bulk of the structure is designed to drop.
  o Predicted debris area - the area beyond the designed area in which it is predicted that the remainder of the debris from the structure could come to rest.
  o Buffer area - the area between the predicted debris area and the exclusion zone perimeter.

Since explosives are not used as common practice in general construction work, advice should be sought from a competent person prior to using any explosives on site.

Applicable Legislation / Standards

* SHWW (General Application) Regulations 1993, Part IV: Work Equipment Regulations.
* SHWW (Construction) Regulations, 2001, Part 8: Explosives, Reg.35; Part 12: Demolition, Regs.47-50.
* SHWW (Explosive Atmospheres) Regulations, 2003.
* SHWW Act, 2005.
* *Proposed SHWW (Construction) Regulations, 2006.*
* *Proposed SHWW (General Application) Regulations, 2006.*

* BS 5607: 1988. *Code of Practice for Safe Use of Explosives in the Construction Industry*.
* BS 6187: 2000. *Code of Practice for Demolition*.

Further Information

* Irish Industrial Explosives Ltd. - advice & information.

* HSA: *All Clear* CD-ROM / Video.
* HSE: CIS45: *Establishing Exclusion Zones when Using Explosives in Demolition*.

DERMATITIS

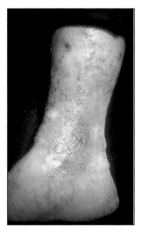

What is dermatitis?

Occupational Dermatitis accounts for more than 50% of all work-related diseases in Ireland. It has long-term consequences for the worker's health and their ability to continue in employment.

There are two types of occupational dermatitis:

* **Irritant Dermatitis** - caused by substances irritating the skin due to constant contact. Common irritants include:
  o Water (wet work).
  o Abrasives.
  o Chemicals.
  o Acids / alkalis / solvents.
  o Detergents.
  o Oils / grease.
* **Allergic Dermatitis** - the worker becomes allergic or sensitised, to the offending substance (allergen). The dermatitis occurs or flares up each time the worker comes in contact with just a small amount of the allergen, even for a short period. The rash may not be confined to the contact area. Common allergens include:
  o Chromate - Cement, primer, paint.
  o Nickel - Jewellery, zips, fasteners.
  o Colophony - Glue, plasticizer, adhesive tape, varnish, polish.
  o Paraphenylenediamine - Dye (clothing, hair), shoes.
  o Rubber chemicals - Tyres, boots, shoes, belts, condoms, gloves.
  o Epoxy resins - Adhesives, plastics, mouldings.
  o Fragrance - Cosmetics / creams, soaps, detergents.

Hazards → Risks

Working with the irritants / allergens listed above leads to **inflammation**, which may lead to **dermatitis**.

Symptoms of inflammation are itching, pain, redness, swelling, and the formation of small blisters or wheals (itchy, red circles with a white centre) on the skin.

Controls → Managing the Risks

Engineering control methods include:

* Enclosure of processes to separate workers from the harmful substances they work with.

* Use of local exhaust systems where toxic substances may escape into the workroom.
* Substitution of non-hazardous for hazardous substances.

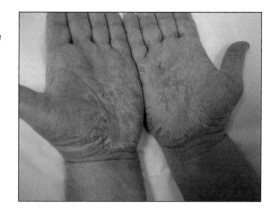

Other control methods include:
* Make an assessment of the risks (employer's responsibility).
* Put in place a system of health surveillance to recognise the early symptoms and train employees to recognise the symptoms.
* Ensure that manufacturer's MSDSs are available.
* Ensure that all necessary PPE is selected by a competent person and is provided to all employees.
* Adhere to manufacturers' specifications for all PPE.
" Use barrier creams as substitutes for protective clothing, especially when gloves or sleeves cannot be used safely.
* Provide advice on personal hygiene, including hand-washing.
* Avoid handling any chemicals, if cuts or scrapes are present on hands or forearms.
* Ensure good house-keeping at all times, including:
  o Proper storage of substances.
  o Frequent disposal of waste.
  o Prompt removal of spills.
  o Maintenance of equipment to keep it free of dust, dirt and drippings.

Applicable Legislation

* SHWW (General Application) Regulations, 1993.
* SHWW (Chemical Agents) Regulations, 2001.
* SHWW (Construction) Regulations, 2001.
* SHWW Act, 2005.
* *Proposed SHWW (Construction) Regulations, 2006.*
* *Proposed SHWW (General Application) Regulations, 2006.*

Further Information

* HSE: *Medical Aspects of Occupational Skin Disease.* Guidance Notes MS24.
* HSE: *Preventing Dermatitis at Work: Advice for Employers and Employees.*

DIESEL FUMES

See also: Machinery - Refuelling

What are diesel fumes?

Diesel engine exhaust emissions / fumes are a mixture of gases, vapours and liquid aerosols. They are a known carcinogen that cause lung cancer and can accumulate in poorly ventilated spaces where engines are operating.

Hazards → Risks

Inhalation of fumes		**Irritation to upper respiratory system - coughing, chestiness, breathlessness Cancer**
Contact with:		
diesel fuel (hot)		**Burns**
diesel fuel (cold)	→	**Dermatitis**

Controls → Managing the Risks

* Install workplace air extraction fans and tail-pipe exhaust extraction systems and filters, as appropriate.
* Use catalytic convectors.

General

* Always turn off engines when not required.
* Keep all doors and windows open, where practicable.
* Use job rotation to minimise exposure.
* Wear suitable PPE at all times and provide workers with training in its correct use and maintenance.
* No smoking or eating in areas where diesel fumes are likely to occur.
* Avoid skin contact with cold or hot diesel fuel or oil.
* Provide all workers with information on the risks of exposure to diesel fumes.
* Report all faults / defects in any controls measures to site management immediately.

Applicable Legislation

* SHWW (General Application) Regulations, 1993.
* SHWW (Chemical Agents) Regulations, 2001.
* SHWW (Construction) Regulations 2001, Part 9: Dangerous or Unhealthy Atmospheres, Regs.36-38.

★ SHWW Act, 2005.

★ *Proposed SHWW (Construction) Regulations, 2006.*

★ *Proposed SHWW (General Application) Regulations, 2006.*

Further Information

★ International Agency for Research on Cancer (IARC) directory of carcinogenic substances - IARC Website.

NOTES

DUMPERS / DUMP TRUCKS

See also   Machinery - General

Dumpers cause 33% of construction traffic accidents

Hazards → Risks

Inadequately maintained braking systems
Driver error, due to lack of
   experience / training
Unstable / rough surfaces
Over-loading

Over-turning
Collisions
Accidents
Injury / Death

Controls → Managing the Risks

✱ Ensure that dumpers are suitable for the tasks intended and that they are maintained per the manufacturer's specifications.

✱ Under no circumstances use machines for work for which they were not designed.

✱ Carry out all operations as per the operator's manual.

✱ Allow only authorised / competent operatives to drive and operate dumpers.

✱ Allow only trained / competent banksmen (CSCS ticket) to give signals to dumper drivers.

✱ Ensure all machinery is fitted with rollover protection / safety frame.

✱ Position stop blocks at a safe distance from the edges of excavations, pits, spoil heaps, etc.

✱ Support edges of roadways and tipping points, where necessary.

✱ Provide physical protection, where necessary to prevent vehicles running off the roadway.

✱ Adhere to site speed limits and conform to site traffic control plans.

✱ Report all damaged / faulty dumpers to the Site Manager / Safety Officer immediately and remove them from work areas.

✱ Remove the keys to dumpers when not in use and park them so as not to present a danger to anyone.

✱ Secure all dumpers (within locked compound / fenced area) when left unattended, to prevent authorised use, especially at the end of the working day.

Drivers should not:
✱ Drive on gradients steeper than those specified as safe in the manufacturer's instructions.

✱ Carry passengers, unless purpose-built seats are provided.

✱ Over-load dumpers.

* Alter tyre pressures outside manufacturer's specifications.
* Use mobile phones while driving dumper.
* Operate dumpers while under the influence of alcohol / drugs, including prescribed drugs.
* Wear loose clothing / jewellery / belts, etc.

Drivers should always:
* Hold valid driving licences and CSCS cards.
* Read the manufacturer's instruction book.
* Understand the differences in performance when loaded / unloaded.
* Check tyres / brakes before operating dumpers.
* Wear appropriate PPE and seat belts.
* Keep to designed vehicle routes and follow site rules.
* Load only on level ground with the parking brake on.
* Ensure loads are evenly distributed and do not obscure visibility.
* Use proper towing pins with jump-out restraints.
* Adhere to site traffic plans and site speed limits.

Applicable Legislation

* SHWW (General Application) Regulations 1993, Part IV: Work Equipment Regulations.
* SHWW (Construction) Regulations, 2001, Part 11: Transport, Earthmoving & Materials Handling Machinery, Regs.41-46.
* SHWW Act, 2005.
* *Proposed SHWW (Construction) Regulations, 2006.*
* *Proposed SHWW (General Application) Regulations, 2006.*

Training

* FÁS CSCS Site Dumper operation.
* FÁS CSCS Articulated Dumper operation.
* FÁS CSCS Slinging / Signalling.

Further Information

* HSE: *Construction Site Transport Safety: Safe Use of Compact Dumpers.* CIS No.52.

DUST

See also   Asthma / PPE / Silica

Hazards → Risks

Cutting concrete / blocks, etc
Sanding   →   Irritation of eyes /
Dry weather periods   ears / throat
Occupational asthma / silicosis

Controls → Managing the Risks

✱ Where possible, minimise all dust generation during operations.

✱ Wet and damp down areas, as necessary, to reduce dispersion of dust.

✱ Where dust is a problem, limit site traffic movement in the area.

✱ Segregate or reduce the numbers of workers exposed.

✱ Wear suitable RPE / PPE, especially where there is a risk from silica dust during scabbling operations.

✱ Monitor dust levels, if a problem arises.

✱ Report a worker's concern that they may be suffering from occupational asthma / dermatitis immediately to the Safety Officer.

✱ Provide medical examinations, if necessary.

CONTROL SYSTEMS

Wet Methods:
Use wet systems on saws powered by combustion engines or compressed air - this involves spraying water onto the rotating cutting disk to reduce dust emissions, using:

> Portable pressurised tank system - supplied by most major plant hire companies.

> Mains water system - essentially the same as the tank, except the water source is from mains supply through a hose to the two water jets.

Local Exhaust Ventilation:
Use the saw's guard as a type of high velocity hood. The guard is connected to an industrial vacuum cleaner, which provides sufficient exhaust ventilation to capture the majority of dust emitted during the cutting operation. Guards with adjustable inner sleeves are preferable, since they maximise enclosure and can be adjusted to accommodate different depths of cut.

Applicable Legislation

* Factory Act, 1955, ss.38, 58.
* Safety in Industry Act, 1980, ss.20, 21 (amends Factory Act, 1955).
* SHWW (General Application) Regulations, 1993, Fifth Schedule (5).
* SHWW (Construction) Regulations, 2001.
* SHWW Act, 2005.
* *Proposed SHWW (Construction) Regulations, 2006.*
* *Proposed SHWW (General Application) Regulations, 2006.*

A dusty atmosphere is hard to see in

Further Information

* HSA: *Occupational Asthma. Save your Breath! An Employees' Guide.*
* HSA: *Occupational Asthma. Save your Breath! An Employers' Guide.*
* HSA: *Occupational Asthma. Save your Breath! Medical Guidance Notes.*
* HSE: *Dust Control on Concrete Cutting Saws Used in the Construction Industry*, CIS54.
* HSE: *Silica*, CIS36.

NOTES

ELECTRICITY

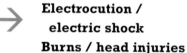

See also: Generators

Hazards → Risks

Operating / servicing / installing /
  removing electrical machinery
  or overhead / underground   →   **Electrocution /
  power lines   electric shock**
Work close to electricity lines   **Burns / head injuries
                                           Unconsciousness / death**

Controls → Managing the Risks

✳ Contact the ESB, or other electrical utility company, for drawings / advice on the position of underground services before any electrical work starts on site.

✳ Request that either power lines are diverted away from the work zone or, if possible, are temporarily switched off to allow work to proceed safely (**main contractor's responsibility**).

✳ Allow only authorised workers, competent to prevent danger, to undertake repairs, servicing, removal, installation and operation of electrical appliances.

✳ Fit all sockets / outlets feeding portable / domestic / water service appliances with RCD protection.

✳ Mark every switch / circuit breaker or control device clearly to indicate 'on' and 'off'.

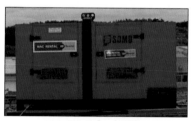

✳ Ensure all working / inspection lights are 25 / 50 volts DC.

✳ Ensure that all electrical panels and distribution boards are suitably identified, properly secured and signed, where necessary, to prevent danger.

✳ Use correct fuses. Do not over-load electrical sockets. Do not use adaptors. Avoid use of extension leads, where possible; if used, place underground or above head height, to avoid mechanical damage.

✳ Erect suitable and sufficient warning signs for all work operations close to power lines.

**Transformer (top);
generator (middle);
portable generator
(bottom)**

Underground Cables / Utilities

* Always assume that underground cables are present, until confirmed otherwise.

* Consult the ESB or other electrical utility company / Safety File / appropriate authority for the location / depth / voltage of cables before any work starts.

* Obtain a drawing and ensure that the ground is surveyed by a competent person, and is marked out to identify the position of all services.

* Before digging, ensure that the ground is scanned with a suitable detector by a competent person to verify the position of any underground services, and mark any differences or variances on the drawing.

Using a cable detector to locate underground electricity lines

* Where a doubt exists as the exact location of a cable / utility, locate it by hand-digging.

* Do not use electrical tools / machinery closer than 0.5m to a known / exposed cable location.

* Where cables / services have been uncovered and remain visible or insufficiently backfilled, erect suitable barriers around the area to make it secure and to prevent unauthorised entry.

* Fit all electric cable in red impact-resistant ducting.

> **Note: Request a representative from the ESB or electrical utility company to be on-site for advice when working close to, or digging close to underground services. Install all electrical cables in accordance with ESB guidelines. Mark the location on an 'as-built' drawing and include the drawing in the Safety File.**

Overhead lines

* Erect warning 'goal posts' at a safe distance on either side of the lines. Ensure that any plant / machinery /vehicles required to pass the goal posts only accesses under the lines via the goal posts. Alternatively, provide suitable warnings and suspended protections where plant / machinery / vehicles have to pass under the lines.

* Create exclusion zones in areas that are unsafe for machinery.

* Do not allow any tipping vehicles or tipping operations close to, or under, overhead lines.

* Guard all exposed lengths of the overhead lines from unapproved access.

* Only use non-conductive ladders near / close to overhead cables.

Take care working near overhead electricity lines - use "goalposts" to avoid contact

Where there will be no work of passage of plant under a over head line:

* A barrier should run parallel to the line - this may be fixed post fencing or steel drums filled with rubble, spaced 1.5 metres apart.
* If cranes are in use, a line of bunting at a height of 3 metres should supplement the barriers.
* Danger notices, stating 'Danger Live Overhead Line', should be spaced at intervals.

High voltage = 1,000 volts AC (alternating current), or 1,500 volts DC (direct current).

Where plant must pass under a live overhead line, the following dimensions must be adhered to:

* Height of goal posts: As advised by ESB.
* Width of goal posts: Max. 10 metres.
* Height of bunting: 3 metres.
* Distance between steel drums: 1.5 metres.
* Distance between danger notices: 20 metres.
* Horizontal distance for barrier to outside conductor on line: 6 metres min. for LV, 10kV, 20kV, 38kV; 10 metres min. for 110kV, 220kV, 400kv.

Avoid using extension leads, as above

Applicable Legislation / Code of Practice

* SHWW (General Applications) Regulations, 1993, Part VIII: Electricity, Regs.33-53.
* SHWW (Construction) Regulations, 2001.
* SHWW Act, 2005.
* *Proposed SHWW (Construction) Regulations, 2006.*
* *Proposed SHWW (General Application) Regulations, 2006, Part VII: Electricity.*
* HSA: *Code of Practice for Avoiding Danger from Underground Services.*

Training / Certification

* National Rules for Electrical Installations.
* Members of Register of Electrical Contractors of Ireland (RECI).

Further Information

* ESB: *The Management of Electrical Safety at Work.* ET 206 - ETCI.

Working at height - p.224

EXCAVATIONS / EARTHWORKS

See also: Excavators / Tunnelling

> **The 4th biggest killer on Irish construction sites – 2 fatalities every year**

What are excavations?

Excavations and trenches greater than 1.25m deep can cause serious accidents from the collapse of their sides, resulting in the burial or crushing of workers inside the excavation.

Key excavation terms include:

* **Shoring** gives temporary support to the wall of a trench by placing sheeting along its walls, with sufficient props both vertically and horizontally to support the length of the excavation exposed.

* **Battering-back** requires the sides of the trench to be sloped back to a 35-45 degree angle to make the sides stable and to prevent collapse.

* **Trench boxes** are proprietary support systems, which can be put in place without requiring people to enter the excavation. Once in place, these allow safe access and work from inside the trench box.

* **Chock / stop blocks** are blocks / structures that prevent a vehicle from approaching too close to the side of an excavation, which could cause the sides of the excavation to collapse or the vehicle to roll into the excavation.

Hazards → Risks

Unprotected edges
Collapse of sides
Machinery / materials falling
   over edges
Contact with underground
   services

Fractures / crush injuries / entrapment / death
Electrocution
Gas inhalation, leading to unconsciousness / death

Controls → Managing the Risks

* Before any excavation operation begins, ensure that the site is inspected and surveyed to ascertain the condition of structures, ground and services, including overhead / underground lines and gas pipes, and that digging operations will not cause other structures to become unstable or collapse.

* Where required, use underpinning and propping to stabilise all adjacent structures that may be affected by excavation operation.

* Ensure that all excavations (more than two metres deep), earthworks, etc are inspected / controlled by a competent person as per Regulations.

* Ensure that CR 9 forms are completed weekly by a competent person and kept on file on-site.

* Provide safe access and egress into the excavation.

* Where necessary, batter-back unstable faces to a safe angle (45 degrees), before work on the construction of basement walls starts.

* Ensure that all shuttering / supports / buttresses, etc. are of sufficient strength and securely supported / strutted / braced to prevent collapse and that they are erected and dismantled only by a competent person under the supervision of a second competent person.

* Use trench boxes to support excavation and to allow safe work inside the trench box. Where necessary, use shoring.

* Ensure that all workers wear approved safety helmets.

* Where required, place suitable barriers, covers and signs around the excavation to forewarn of the danger.

* Inspect the ground area close to the excavation to ensure it is capable of taking the weight of plant and equipment.

* Do not operate machinery too close to the edges of any excavations.

" Put in place stop / chock blocks to prevent machinery driving too close to the edge of the excavation.

An excavation battered-back (top) and a graded excavation (bottom)

* Store all material removed from the excavation at least the same distance away from the edges as the depth of the excavation, to prevent debris falling back in.

* Ensure that there is good communications / co-ordination between the machine drivers and other workers during all digging operations.

* Where combustion engines are used in an excavation, ensure that the exhaust pipes extend clear of the excavation and that the work area is properly ventilated.

A competent person must supervise the installation, alteration or removal of excavation support.

A competent person must inspect excavations:

* At the start of each shift, before work begins.

* After any event likely to have affected the strength or stability of the excavation.

* After any accidental fall of rock, earth or other material.

WARNING
Don't get buried alive! One cubic metre of soil weighs 1 tonne+. This is enough to kill you!

A written report should be made after each inspection (CR9 form).

Before work starts, consider:

* Equipment needed - trench sheets, props, baulks, etc - and its availability.
* Previous use of site / location of existing buildings.
* Amount of storage and working space required.
* Level of water table and type of soil.
* Ground contamination.
* Storage and disposal of excavated material.
* Suitable method for temporary support and prevention of collapse / falls.
* Emergency arrangements.

Applicable Legislation

* SHWW (General Applications) Regulations, 1993.
* SHWW (Confined Spaces) Regulations, 2001.
* SHWW (Construction) Regulations, 2001, Part V: Regs.21-26.
* SHWW Act, 2005.
* *Proposed SHWW (Construction) Regulations, 2006.*
* *Proposed SHWW (General Application) Regulations, 2006.*

Required Documentation

* HSA Form CR 9: Excavations, Shafts, Earthworks, Underground Works or Tunnels - Report of thorough examination.

Further Information

* HSA: *A Guide to Safety in Excavations*.
* HSA: *Safe System of Work Plan (SSWP) Ground Works Form*.
* HSE: *Safety in Excavations*, CIS8 (rev.1).

Quick Checklist

Are controls in place to prevent:
> Collapse of the sides?
> Materials falling into excavation?
> People and vehicles falling into excavation?
> People being struck by plant?
> Undermining nearby structures?
> Contact with underground services?
> Unauthorised access / egress?
> Fumes?
> Accidents to members of the public?

EXCAVATORS (180° & 360°)

See also   Excavations & Earthworks / Lifting Equipment
& Operations / Machinery - General

Hazards → Risks

Testing / servicing machinery
   on uneven / sloped surfaces
Untrained staff
Unauthorised use /
   carrying of passengers
Servicing elevated parts /
   components
Pressure failure

→

**Overturning / crushing
Amputation /
   entanglement
Serious injury / death**

Controls → Managing the Risks

* Operate and maintain all excavators as per operator's handbook. Keep a copy of the handbook in the cab of the excavator at all times.
* Ensure that drivers remain in the safety cab (and wear safety belts, where fitted) in the event of an excavator overturning and that they do not attempt to jump clear.
* Allow only trained / competent (valid CSCS cards) workers to drive excavators.
* Allow only trained / competent banksmen (valid CSCS card) to give signals to excavator drivers.
* Do not allow any person under 18 to operate an excavator or to give signals to the driver.
* Do not operate excavators while under the influence of alcohol / drugs, including prescribed drugs.
* Do not use mobile phones while operating excavators.
* Do not wear loose clothing / jewellery / belts, etc. while operating excavators.
* No unauthorised riding on excavators - one seat, one person.

* Ensure that all excavators adhere to site speed limits and conform to site traffic control plans.
* No bypassing or interfering with safety devices on the excavators or buckets.
* Use hydraulics only within their design limit / as per supplier's instructions.
* Before adjusting / repairing / maintaining / leaving the excavator, always turn off the engine.
* Under no circumstances rely on the hydraulics to maintain raised equipment while working underneath.
* Keep other workers clear of the excavator boom swing area.
* Lower the boom / bucket when not in use.
* Adhere strictly to the SWL on the arm of the excavator.
* Secure excavator buckets to a quick hitch - using a locking pin to prevent the inadvertent dropping of the bucket.
* Secure all excavators when left unattended, to prevent authorised use, especially at the end of the working day.

Applicable Legislation

* SHWW (General Application) Regulations, 1993, Part IV: Work Equipment.
* SHWW (Construction) Regulations, 2001, Part 11: Transport, Earthmoving & Materials Handling Machinery, Regs.41-46; Part 14: Lifting Appliances: Excavators Used as Cranes, Reg.100; Seventh Schedule: Minimum Requirements for an Excavator or Loader Used as a Crane.
* SHWW Act, 2005.
* *Proposed SHWW (Construction) Regulations, 2006.*
* *Proposed SHWW (General Application) Regulations, 2006, Part III: Use of Work Equipment.*

Required Documentation

* CR2 - Excavator or Loader Used as Crane - Certificate of SWL.

A "rubber duck" is an excavator with wheels instead of tracks

Training / Certification

* FÁS CSCS 180° Excavator operation.
* FÁS CSCS 360° Excavator operation.

Programme Content includes:
* Safely operate and control a 360° hydraulic excavator machine.
* Excavate to line and level for land drainage, land reclamation, civil engineering, bulk earthworks, pipe-laying and rock-breaking.
* Check, inspect and maintain a 360° excavator for efficient use under construction site conditions.
* Complete site clearance and / or reinstatement.
* Load a dump truck.
* Load / un-load a 360° excavating machine on / off a low-loader.

Further Information

* Operator's manual and manufacturer's guidelines.

NOTES

FALL PROTECTION EQUIPMENT

See also: MEWPs / Roof Work / Scaffolding / Working at Height

What is fall protection?

There are two types of fall protection:

* Work restraint (fall restraint) system, which stops a person from falling.
* Fall arrest system, which stops the person after they have fallen.

Hazards → Risks

Incorrect equipment for task
Worn / damaged lanyards   →   **Falls, resulting in**
  and / or harness   **injuries or death**
Harness failure

Untrained operatives
Rescue attempts   →   **Suspension trauma**

Controls → Managing the Risk

* Ensure that fall protection equipment is selected by a competent and trained person with the advice of suppliers.
* Ensure that the type of fall protection selected is based on the results of a Risk Assessment.
* Ensure that a suitable rescue procedure is planned before harnesses are used.
* Always establish the minimum clearance height for the proposed equipment: A typical fall arrest system with a full body harness, 2.0m lanyard and shock-absorbing device requires over 5m clearance height to deploy and arrest a fall.
* Ensure that the lanyard length (of both fixed length and retractable systems) is carefully selected by a competent person.
* Ensure that there are no projections that a person could hit during a fall.
* Do not allow working at heights and use of fall protection equipment in extreme weather conditions / insufficient lighting.
* Never allow workers to work alone using fall protection equipment.

Suspension Trauma

Workers who are immobile while suspended in a harness during post-fall suspension can suffer suspension trauma, resulting in loss of consciousness and death.

* Ensure that trained rescuers are available when work restraint / fall protection systems are being used and that they are familiar with the recognition, and treatment, of suspension trauma.
* Ensure that workers are instructed in the use of the harness, lanyard, rescue equipment and in the inspection, maintenance and storage of fall protection PPE.

A safety harness in use

Applicable Legislation / Standards

* SHWW (General Application) Regulations, 1993.
* SHWW (Construction) Regulations 2001: Part 13, Reg.79(1-4).
* SHWW Act, 2005.
* *Proposed SHWW (Construction) Regulations, 2006.*
* *Proposed SHWW (General Application) Regulations, 2006, Part XV: Work at Height.*

* BS / EN 795:1997. PPE against falls from a height: Anchor devices.
* BS / EN 345:2002. PPE against falls from a height: Lanyards.
* BS / EN 355:2002. PPE against falls from a height: Energy absorbers.
* BS / EN 358:2002. PPE against falls from a height: Belts for work positioning & restraint & work position lanyards.
* BS / EN 360:2002. PPE against falls from a height: Retractable type fall arrestors.
* BS / EN 361:2002. PPE against falls from a height: Fully body harnesses.
* BS / EN 363:2002. PPE against falls from a height: Fall arrest systems.

Further Information

* HSA: *Guide to Lanyards.*
* HSE: *Suspension Trauma*, Research Report.

FALLING OBJECTS

See also   House-keeping / Public Safety
/ Roof Work / Scaffolding /
Working at Height

> 16% of fatalities in 2004 were workers struck by falling objects

Hazards → Risks

Falling objects / persons   →   Fractures / lacerations /
contusions / death

Controls → Managing the Risks

* Do not carry out work while the area underneath is occupied. Mark out a dedicated exclusion zone or provide overhead protection.
* Fit toe-boards to all scaffolding and ledges.
* Fit safety nets, where appropriate.
* Ensure that all workers wear hard hats.
* Stack materials / equipment in such a way as to prevent collapsing / overturning.
* Store all materials in designated areas - for example, loading bays.
* Do not clutter working platforms with stored materials or debris.
* Adhere to a 'clean as you go' policy.
* Use chutes where possible for discarding materials from heights. Never throw objects or materials from heights.
* Use sheeting / nettings / hoarding to enclose scaffoldings / roofs, especially when working close to the public.
* Use appropriate / suitable tool pouches at all times.
* Remove or tie down all loose materials in bad weather.
* Cordon off areas under slewing cranes.

Toe-boards (top) and safety netting (bottom) can help to prevent accidents due to falling objects, such as hammers (below)

Applicable Legislation / Standards

* SHWW (General Application) Regulations, 1993.
* SHWW (Construction) Regulations 2001, Part 4: Reg.16(1-4); Part 3: Reg.79(1-4).

* SHWW Act, 2005.
* *Proposed SHWW (Construction) Regulations, 2006.*
* *Proposed SHWW (General Application) Regulations, 2006, Part XV: Work at Height.*

DANGER
Falling
objects

Safety netting

Safety netting is a relatively new method of fall prevention, but is effective and efficient in combating falls and falling objects.

* EN 1263, Part 1 is the Standard for the manufacture of safety nets.
* EN 1263, Part 2 is the Standard for the installation of safety nets.

Soft landing systems

The Soft Landing System has been designed for use principally inside a building during construction, where the bags will be enclosed by walls or partitions.

The bags are most commonly used at ground floor level while the joists are being boarded. This ensures that the carpenter will have a soft landing in the event of a leading edge fall while affixing the boards across the joists. On first or subsequent floors, the system may be used to protect workers while trusses are being installed, by installing bags on the covered joists.

Putting a soft landing system in place

NOTES

FENCING

See also: Site Security

Hazards → Risks

Erection / dismantling of post
and wire fencing
Treated timber / line wire /
barbed wire

→ **Lacerations / injury**

Controls → Managing the Risks

* Check the proposed fence line for underground hazards - cables / water / gas.
* When stake driving, do not support the stake by hand - use a stakeholder.
* When using a maul / mallet, ensure that no one is close in line with the swing.
* Where possible, use treated timber and check that the preservative is dry.
* Wear chemical-resistant gloves to handle timber still wet with preservative.
* Use dispensers when unrolling line wires to avoid kinking / twisting.
* Ensure that the wire is secured firmly on the dispenser.
* When using ratchets, ensure that the wire always has at least two full turns on the ratchet barrel.
* Ensure wire strainers are securely attached and anchored before tensioning.
* Do not stand on, or astride, wire while it is being tensioned.
* Never over-tension by using extra leverage.
* To avoid recoil, always ensure that the exposed ends of wire are secured.
* Always secure wire on each side of the cutting point before cutting.
* Exercise care to avoid spiking hands on

Barbed wire coil (top) and different kinds of fencing (above)

loose ends.

* Always wear protective gloves when handling barbed wire.
* Take care to avoid breakage and recoil.
* When dispensing barbed wire, keep it taut.
* Ensure that boundary fencing is not be a danger in itself, and that it is designed not to cause injury to trespassers.

A proprietary fencing / handrail system

Applicable Legislation

* SHWW (General Application) Regulations, 1993.
* Occupiers' Liability Act, 1995.
* SHWW (Construction) Regulations, 2001.
* SHWW Act, 2005.
* *Proposed SHWW (Construction) Regulations, 2006.*
* *Proposed SHWW (General Application) Regulations, 2006.*

Further Information

* Manufacturers' / suppliers' guidelines and information.

NOTES

FIRE

See also: House-keeping / Machinery - Refuelling / Welfare

Hazards → Risks

Fire   →   **Burns / scalds**
                     Smoke inhalation / suffocation
                     Death

Types of Fires / Appropriate Fire Extinguishers

Fire Code	Fuel
A	Wood, paper, cloth, plastics
B	Flammable liquids
C	Electrical
D	Metals

Extinguisher type	Use for
Water	A
CO_2	B / C
Powder	A / B / C
Foam	A / B

WATER
CO_2
DRY POWDER
FOAM

Controls → Managing the Risks

* Maintain all areas free from fire hazards, as far as is reasonably practical.
* Keep all areas (inside and out) clear of any accumulation of rubbish or combustible materials.
* Store solvents / cleaners correctly and in their correct containers.
* Train all workers in the recognition of the causes of fire, the correct type of fire extinguisher to be used and how to raise the alarm.
* Ensure all workers are familiar with at least two escape routes from their work area, also with the fire extinguishers in their area.
* Keep all passage-ways / escape routes clear.
* Do not over-load electrical equipment.
* Maintain / operate all appliances as per the manufacturer's instructions.

* Ensure that all workers familiarise themselves with the emergency evacuation plan.
* Ensure that escape routes and fire points are signed, kept clear at all times, and equipped with emergency lighting.
* Select and sign assembly areas.
* Develop and rehearse emergency fire procedures. Carry out emergency evacuation / fire drills at least twice a year.
* Appoint a Fire Warden.
* Ensure that a Permit to Work system is in place for all "hot" work.

Applicable Legislation / Code of Practice

* Factory Act, 1955, ss.45, 46, 47, 48.
* Fire Services Act, 1981.
* Fire Safety in Places of Assembly (Ease of Escape) Regulations, 1985.
* SHWW (General Application) Regulations 1993, Reg.17; Second, Third & Fourth Schedule.
* Building Regulations, 1997.
* SHWW (Construction) Regulations, 2001, Part 4: Reg. 17; Fourth Schedule.
* SHWW Act, 2005.
* *Proposed SHWW (Construction) Regulations, 2006.*
* *Proposed SHWW (General Application) Regulations, 2006.*

* Department of Environment: *Codes of Practice for Fire Safety.*

Further Information

* British Home Office: *British Fire Precautions (Workplace) Regulations, 1997.*

FIRST AID

See also: Accidents

What is First Aid?

First Aid is defined as 'treatment for the purpose of preserving life, or minimising the consequences of injury or illness, where a person requires medical treatment in a case of a minor injury which would otherwise receive no treatment or which does not require medical treatment'.

Hazards → Risks

Incorrect treatment   **Further injuries / death**

Controls → Managing the Risks

* Provide trained Occupational First Aiders on each construction site, as required.
* Ensure that First Aiders' certification is by a recognised Occupational First Aid Instructor.
* Provide and maintain adequate / appropriate First Aid equipment.
* Report all accidents to the Site Manager / Safety Officer immediately.
* Complete the on-site Accident Report Book for each incident (even cut fingers).
* Do not administer drugs or medications (not even over-the-counter tablets, such as Disprin).
* Keep records of all accidents / reportable accidents / dangerous occurrences.

Applicable Legislation

* SHWW (General Application) Regulations 1993, Part IX: Regs.54-57; Part X: Regs.58-63.
* SHWW (Construction) Regulations, 2001.
* S.I. No.5 of 2003, Reg.56(1)(b).
* *Proposed SHWW (General Application) Regulations, 2006, Part VIII: First Aid; Part IX: Reporting of Occupational Accidents & Dangerous Occurrences.*

Site First Aid Kits - suggested contents:
> Adhesive plasters.
> Sterile eye-pads.
> Triangular bandages.
> Sterile wound dressings.
> Anti-septic wipes.
> Cutting shears.
> Latex gloves.
> Eye-wash.

Training Requirements

* 1993 Regulations, Reg.54 - Occupational First Aider.

Approved Training Bodies or Organisations:
* Civil Defence School.
* Irish Red Cross Society.
* National Ambulance Training School.
* Order of Malta Ambulance Corps.
* St. John Ambulance Brigade of Ireland.
* Any qualified First Aid instructor registered with the National Ambulance Training School.

Further Information

* HSA: *Guidelines on First Aid*.
* HSA: *Guidelines to the Safety, Health & Welfare at Work Act, 1989 & the Safety, Health and Welfare at Work (General Application) Regulations, 1993, Part IX: First Aid* (pp.156-164).
* HSE: *Basic Advice on First Aid at Work*.
* HSE: *First Aid at Work: Your Questions Answered*.
* National Ambulance Training School: *Register of Approved First Aid Instructors*.

A First Aid kit is essential in any site office - ability to perform CPR is essential - by law, there must be an adequate number of qualified First Aiders on site

NOTES

FORKLIFTS

See also: Falling Objects / Teleporters / Working at Height

Hazards → Risks

Loading / unloading /
  transporting
Pedestrians / traffic
Slippery surfaces
Unstable loads
Falling objects

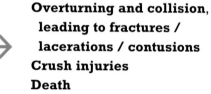

**Overturning and collision,
  leading to fractures /
  lacerations / contusions
Crush injuries
Death**

Controls → Managing the Risks

* Adhere strictly to the manufacturer's handbook and operator's manual instructions.
* Ensure that a competent person inspects the forklift at least once every 14 months and that they complete the necessary documentation.
* Do not allow make-shift repairs or unauthorised use.
* Keep all areas clear of obstructions and substances that contribute to slippery surfaces.
* Allow only authorised trained / qualified workers (over 18) to operate forklifts.
* Keep records of all training and its content.
* Do not carry passengers on forklifts, unless they are designed to do so.
* Use only designated traffic routes - with the utmost care.
* Operate forklifts only on flat, level surfaces.
* Use proper lifting techniques only, at speeds that do not pose a danger.
* Ensure that forks are loaded correctly / not over-loaded.
* Secure all loose material before lifting.
* Ensure that loads are carried as close to the ground as practicable.
* Do not use mobile phones when operating forklifts.
* Secure forklifts when left unattended, to prevent unauthorised use, especially at the end of the working day.

Applicable Legislation / Code of Practice

* SHWW (General Application) Regulations, 1993, Part IV: Work Equipment Regulations.
* SHWW (Construction) Regulations, 2001, Part 11: Transport, Earthmoving & Materials Handling Machinery, Regs. 41-46.

* SHWW Act, 2005.
* *Proposed SHWW (Construction) Regulations, 2006.*
* *Proposed SHWW (General Application) Regulations, 2006.*

* HSA: *Code of Practice for Rider-Operated Lift Trucks: Operator Training & Supplementary Guidance.*

Required Training

All workers who use forklifts should have completed a recognised training course from a reputable training provider.

NOTES

FORMWORK

See also: Concrete & Cement

What is formwork?

Also known as 'falsework', 'temporary work' or 'propping', formwork is any work that is necessary for the safe construction of the permanent works but not for the safety or proper function of the permanent works after completion - for example, a temporary structure used to support a permanent structure during construction until it is self-supporting.

Temporary structures are required in the construction stages to prevent collapse due to overloading of structural components during the building and installation works - for example, the installation of pre-cast slabs and stairs, etc.

The responsible contractor must ensure that the correct number and type of props are installed correctly and that the units are supported as indicated on the construction drawings.

Hazards → Risks

Unauthorised alterations / removal of formwork

Over-loading of formwork

Collapse of structure

Falling materials

Injury / unconsciousness / death

Controls → Managing the Risks

* Appoint a formwork co-ordinator as soon as possible.
* Ensure all formwork follows the designer's instructions.
* Make clear that no worker has permission to interfere, alter or dismantle formwork.
* Ensure that all formwork is strong enough and stable in use.
* Do not use damaged components. Inspect parts regularly for defects / damage.
* Ensure that load-bearing connections, including the use of angle brackets and bolts, are specified by the designer.
* Ensure that bolts are of suitable size and are inserted to the required depth to the designed centre distances.

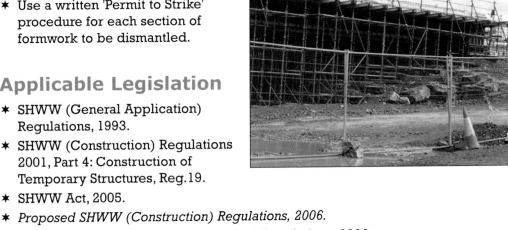

* Ensure that all formwork is inspected and certified as 'ready to use' by a competent person.
* Use a written 'Permit to Strike' procedure for each section of formwork to be dismantled.

Applicable Legislation

* SHWW (General Application) Regulations, 1993.
* SHWW (Construction) Regulations 2001, Part 4: Construction of Temporary Structures, Reg.19.
* SHWW Act, 2005.
* *Proposed SHWW (Construction) Regulations, 2006.*
* *Proposed SHWW (General Application) Regulations, 2006.*

Further Information

* HSE: *Safe Erection, Use & Dismantling of Falsework*, CIS56.

NOTES

GAS & GAS CYLINDERS

See also: Confined Spaces

Hazards → Risks

Inhalation of toxic fumes	→	**Nausea**
		Asphyxia / suffocation
		Unconsciousness
		Death
Manual handling	→	**Back injury**

Controls → Managing the Risks

* Before starting work, contact Bord Gáis Éireann or other gas utility company and request drawings showing the location of underground gas services.
* Request that either gas services be diverted away from the work zone or, if possible, they be temporarily cut off to allow work to proceed safely (**main contractor's responsibility**).
* Obtain drawings of the underground services and other related gas services. Survey and mark out the ground or walls to identify the position of services.
* Consult the Safety File for extensions or repairs (existing properties).
" Upon completion of the project, update the Safety File with drawings of the locations of all services.
* Before digging, identify the location of gas services. Do not allow mechanical excavation around or close to underground services. Use hand-digging only to unearth or make visible underground gas services.
* Do not allow sources of ignition in the vicinity of live gas. No smoking.
* If a gas leak is identified, erect warning signs and suitable barriers.

Storage of gas cylinders:
* Store all gas cylinders in a safe place in the open air.
* Store all cylinders in an upright position, properly secured.
* Protect gas cylinders from external heat sources.
* Do not raise or lower cylinders on the forks of a forklift, unless adequate precautions are taken to prevent them falling.
* Do not lift cylinders by their valves.

✴ Take care to ensure that cylinders are not dropped.

Applicable Legislation

✴ SHWW (General Application) Regulations, 1993.
✴ SHWW (Construction) Regulations 2001, Part 9: Dangerous or Unhealthy Atmospheres, Regs.36-38.
✴ SHWW Act, 2005.
✴ *Proposed SHWW (Construction) Regulations, 2006.*
✴ *Proposed SHWW (General Application) Regulations, 2006.*

A confined spaces gas detector

Further Information

✴ Bord Gáis Éireann.
✴ BOC Gases - MSDS sheets / Gas Safety Awareness training / safety videos.

NOTES

GENERATORS

The "silent killer"

See also: Electricity / Hand & Power
           Tools

Hazards → Risks

Toxic fumes, from
  engines operating /   →
  running in enclosed areas

Nausea
Asphyxia
Unconsciousness
Death

Controls → Managing the Risks

* Adhere strictly to manufacturer's instructions.
* Allow only competent / qualified staff to supervise operations.
* Store diesel properly.
* Do not operate generators indoors, where possible.
* Where combustion engines must operate indoors or in confined spaces, such as workshops / deep excavations, etc, put in place adequate and suitable exhaust systems to ventilate the workshop / excavation so as not to endanger workers.
* Do not over-load generators.
* Ensure that generators are selected for use by a competent person.
* Limit access to generators to competent persons only.

Applicable Legislation

* SHWW (General Application) Regulations, 1993, Part IV: Work Equipment.
* SHWW (Construction) Regulations 2001, Part 9: Dangerous or Unhealthy Atmospheres, Regs.36-38.
* SHWW Act, 2005.
* *Proposed SHWW (Construction) Regulations, 2006.*
* *Proposed SHWW (General Application) Regulations, 2006.*

Further Information

* HSA: *Carbon Monoxide, The Silent Killer.*

Blocks / Bricks - see p. 56

HAND & POWER TOOLS

See also: **Abrasive Wheels / Augers & Drills / Cartridge-operated Tools / Chainsaws / Concrete & Cement / Generators / Saws / Vibration**

Hazards → Risks

Electrocution **Shock / fractures / lacerations / contusions / fragments**

Vibration

Controls → Managing the Risks

* Operate and maintain all tools as per the manufacturer's instructions.
* Service all tools as required and keep a record of the servicing.
* Use only suitable / appropriate power tools for each task - size, condition, voltage, etc.
* Allow only qualified / competent operatives to operate power / electrical tools.

* Ensure that all portable and electrically-powered tools on site are supplied at 110V.
* Never carry tools by cords and never pull cords to disconnect.
* Never use power tools with leads that use insulating tape.
* Check all tools for faults and damage before use.
* Store all tools correctly after use.

* Report all defects / faults immediately to the Site Manager / Safety Officer.
* Remove damaged, defective, worn or suspect tools from service and mark them 'out of service' until repaired.
* Allow only competent persons (for example, qualified electricians) to carry out repairs / adjustments, and then only with the appliance / tool disconnected from the power source.
* Ensure that all tools with rotating shafts and components have suitable guards fitted.
* Do not tamper with guards / safety devices.
* Ensure that tools that generate dust or fumes are fitted with appropriate extraction or wetting aids and, where possible, are only used outdoors.

* Avoid trailing electric cables and extensions leads, where possible. Use only correctly rated leads (fully uncoiled), elevated above head level, where absolutely necessary.

* Where mains electricity is used as a power source, ensure that a RCD (rated at 30m Amps, with no time delay) is fitted by a competent person.

* Wear additional PPE when operating power tools - goggles / eye protection / gloves, as necessary.

Sanders

* Check belts before use for damage / tears. Replace when necessary.
* Check the direction of the rotation.
* Ensure that all work pieces are adequately supported and that all stops are made secure.
* Ensure that nip-guards are in place / secure / serviceable.

5 Golden Rules

1. Keep tools in good condition with regular maintenance.
2. Use the right tool for the job.
3. Examine tools for damage before use.
4. Operate tools according to the manufacturer's instructions.
5. Provide and use the proper PPE.

Applicable Legislation

* SHWW (General Application) Regulations, 1993, Part IV: Work Equipment Regulations.
* SHWW (Construction) Regulations, 2001.
* SHWW Act, 2005.
* *Proposed SHWW (Construction) Regulations, 2006.*
* *Proposed SHWW (General Application) Regulations, 2006.*

Further Information

* Manufacturer's / supplier's instructions.

HARASSMENT

See also: Bullying / Stress

What is harassment?

Examples of behaviour that may be classified as harassment include:

* Sexual / racial / gender / general discrimination.
* Purposely undermining someone / victimisation.
* Humiliation.
* Social exclusion or isolation.
* Intimidation.
* Verbal or physical abuse or threats of abuse.
* Aggressive or obscene language.
* Intrusion by pestering, spying or stalking.

Sexual harassment is defined in the Equality Authority Code by reference to the Employment Equality Act, 1998, s.23, which the Authority summarises by noting that 'sexual harassment includes any act of sexual intimacy, request for sexual favours, and / or other act or conduct, including spoken words, gestures or the production, display or circulation of written words, pictures or other material that is unwelcome and could reasonably be regarded as sexually offensive, humiliating or intimidating'.

Hazards → Risks

Ill-health
Fear   **Bodily harm / stress**
Anxiety / depression   **Absenteeism**

Controls → Managing the Risks

* Make clear that the company will not tolerate harassment under any circumstances.
* Adhere to the company's Policy Statement on Dignity at Work.
* Arrange regular briefings with employees on the signs / effects of harassment, and put in place a confidential reporting system.
* Refer difficulties with the public / other contractors to the Site Manager / PSCS / Safety Adviser.
* Call the Gardaí in serious cases.

Applicable Legislation

* SHWW (General Application) Regulations, 1993.
* S.I. No.72 of 2002. Code of Practice on Sexual Harassment & Harassment at Work (Equality Authority Code).
* Equality Act, 2004.
* SHWW Act, 2005.
* *Proposed SHWW (General Application) Regulations, 2006.*

Further Information

* Equality Authority.

NOTES

HOUSE-KEEPING

**A clean site
is a
SAFE
site**

See also   **Falling Objects / Fire / Office
Work / Slips & Trips**

Hazards → Risks

Access, egress and general
  movement / slippery surfaces /
  poor house-keeping / poor lighting

→

**Slips / trips / falls
Back, hand or head
 injuries**

Controls → Managing the Risks

* Put in place a good house-keeping programme to manage the orderly movement of materials / persons from the point of entry to exit and within the site, and the cleaning of all areas.

* Carry out periodic checks to ensure that no obvious danger exists that might endanger workers.

* Keep all areas clear of obstructions.

* Ensure that all workers / sub-contractors adopt good house-keeping practices and operate a 'clean as you go' policy.

* Ensure that workers report to management any unusual conditions they discover.

* Ensure that all timber is de-nailed, or nails knocked back, before discarding.

* Never drop materials or rubbish from heights. Use chutes for all loose materials or gather rubbish into bundles or containers and lower safely to ground level.

* Place all debris / rubbish in designated skips.

" Make available separate skips for timber / metal / plastic / canteen skips.

* Ensure that canteen skips, or skips containing waste food products, are enclosed or covered.

Applicable Legislation

* SHWW (General Application) Regulations, 1993.

* SHWW (Construction) Regulations, 2001.

* SHWW Act, 2005.

* *Proposed SHWW (Construction) Regulations, 2006.*

**Separate skips for
different types of waste
(top); and tidy storage
(below) contribute to a
safe site**

✴ *Proposed SHWW (General Application) Regulations, 2006.*

Further Information

✴ HSA: *Guide to The Safety, Health & Welfare and Work Act, 1989 & The Safety, Health & Welfare at Work (General Applications) Regulations, 1993, Part III: Workplace.*

NOTES

LADDERS

See also: Working at Height

Hazards → Risks

Falls
→
**Accident / injuries
Back injury / paralysis /
unconsciousness / death**

Controls → Managing the Risks

* Before use, ensure that the ladder is not damaged and is suitable for the task, is secure and cannot slip.
* Do not use make-shift, painted or home-made ladders.
* Always ensure that the ladder is angled to minimise the risk of slipping.
* Set ladders on a firm, level base - do not use make-shift props.
* Do not stand on the top three rungs of ladders.
* Do not secure ladder by its rungs - place the lashing around the stiles, or use ladder ties / clips.
* Do not place ladders against fragile surfaces or fittings. Use bracing boards on openings.
* Do not over-reach when on ladders, or carry heavy loads up / down ladders.
* Climb / descend the ladder facing the ladder and using both hands for security.
* Report ladders showing signs of defects to Site Manager / Safety Officer and remove them from the site.
* Do not drop or throw a ladder.
* Take care that ladders do not come in contact with power cables.
* Do not use aluminium ladders near overhead cables.
* Spread open stepladders securely. Never use a folding stepladder in an unfolded position.
* Remember that ladders are for access only and are never to be used as a working platform.
* As a general rule, ensure that you always have three points of contact with ladder - two feet and one hand.
* As a general rule, incline a ladder at an angle of 1:4.

Safety: ladders tied down (top); and appropriate signage (below)

* Where the ladder is over 3m high, secure it at the top. Ensure that the ladder extends a minimum height (about 1m) above any landing area, unless a suitable scaffold handhold is available.
* Ensure that all ladders to scaffolding are installed by a competent person or scaffolding company and that they are not be tampered with by other workers.

Applicable Legislation

* SHWW (General Application) Regulations, 1993, Part IV: Work Equipment Regulations.
* SHWW (Construction) Regulations, 2001, Part 13: Working at Heights, Regs.72 & 73.
* SHWW Act, 2005.
* *Proposed SHWW (Construction) Regulations, 2006.*
* *Proposed SHWW (General Application) Regulations, 2006, Part XV: Working at Height.*

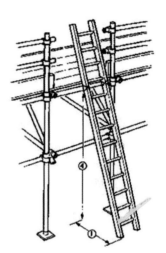

Further Information

British Ladder Manufacturers Association:
* *Ladders & the Work at Heights Regulations.*
* *Risk Assessment Guide - Step-Ladders.*
* *Risk Assessment Guide - Leaning Ladders.*
* *The Ladder Safety Manual.*

NOTES

LASER EQUIPMENT

What is laser equipment?

Surveying equipment is the best example of laser equipment on site, such as the tripod below.

Hazards → Risks

Direct eye-contact with laser → **Eye injury / blindness**
Burns
Electric shock

Controls → Managing the Risks

* Operate and maintain all laser equipment as per the manufacturer's instructions.
* Ensure that no worker looks at the laser beams or directs the beams towards other workers.
* Ensure that laser equipment is serviced / repaired only by the manufacturer.
* Do not use wet batteries / charger.
* Do not cover the battery charger while charging.
* Do not use laser equipment in explosive / damp atmospheres, unless rated suitable / safe by the manufacturer.

Applicable Legislation

* SHWW (General Application) Regulations, 1993, Part IV: Work Equipment Regulations.
* SHWW (Construction) Regulations, 2001.
* SHWW Act, 2005.
* *Proposed SHWW (Construction) Regulations, 2006.*
* *Proposed SHWW (General Application) Regulations, 2006.*

Further Information

* Manufacturer's operational manual and guidelines.

Cranes & Lifting operations - see p. 84 / 136

LEAD

See also: PPE / Pregnant Employees

Hazards → Risks

High-risk activities
(see below) → **Headaches, tiredness, irritability, constipation, nausea, stomach pains
Kidney damage / nerve and brain damage**

High-risk activities include:
* Lead smelting, refining, alloying and casting.
* Lead-acid battery manufacture and breaking.
* Manufacturing lead compounds.
* Manufacturing leaded-glass.
* Manufacturing and using pigments, colours and ceramic glazes.
* Working with metallic lead / alloys containing lead - for example, soldering.
* Some painting of buildings; some spray-painting of vehicles.
* Blast removal and burning of old lead paint.
* Stripping of old lead paint from doors, windows, etc.
* Hot cutting in demolition and dismantling operations.
* Recovering lead from scrap and waste.

Controls → Managing the Risks

* Wear the correct PPE at all times.
* Where the Risk Assessment shows the need, provide adequate RPE.
* Report any damaged / defective ventilation plant or protective equipment to site management immediately.
* No eating, drinking or smoking outside of designated areas.
* Ensure that employees wash their hands and face and scrub their nails before eating, drinking or smoking. Provide showers, where necessary.
* Provide training and information to all employees on the risks to health and the precautions / controls necessary in relation to lead.
* Measure and monitor the level of workers' exposure on a regular basis.
* Provide health surveillance to all employees.

Lead ore, also known as galena

Applicable Legislation

* SHWW (General Application) Regulations, 1993.
* Maternity Protection Act, 1994.
* SHWW (Pregnant Employees, etc) Regulations, 2000.
* SHWW (Construction) Regulations, 2001.
* SHWW Act, 2005.
* *Proposed SHWW (Construction) Regulations, 2006.*
* *Proposed SHWW (General Application) Regulations, 2006, Part XI: Pregnant, Post-Natal and Breastfeeding Employees.*

WARNING

A developing, unborn child is at particular risk from exposure to lead, especially in the early weeks before a pregnancy becomes known.

Further Information

* HSE: *Lead & You*, INDG305.

NOTES

LIFTING EQUIPMENT & OPERATIONS

See also: Chains & Slings / Cranes / Excavators / Machinery - General / Pre-cast Elements & Components / Teleporters / Working at Height

What are lifting equipment and operations?

Lifting equipment means a gear or cable by which a load can be attached to a lifting appliance, and can include chain slings, rope slings, hooks, shackles or eye bolts.

Hazards → Risks

Failure of a load-bearing element
  of the equipment
Over-turning of the equipment
Contact with the load
Loads falling from a height

Collision
Crushing
Other serious injuries
Death

Controls → Managing the Risks

* Ensure that all lifting equipment used on-site is CE-marked, or certified as designed to a relevant recognised international standard, particularly in relation to protection of the operator from falling objects and over-turning.

* Ensure that all new equipment is Risk-Assessed before use, with particular reference to the proposed area of use.

* Ensure all equipment is inspected and certified by a competent person, that all certificates are kept on file on-site and that all equipment is re-tested and re-certified periodically.

* Label defective equipment and take it out of service until repaired and re-certified by a competent person.

* Do not start lifting operations before a Method Statement has been prepared and approved.

* Ensure that operators of all tower cranes, teleporters, mobile cranes, etc are CSCS-certified.

* Allow only certified slingers / signallers to guide lifting operations.

All lifting operations should be guided by an experienced banksman

* Before any lifting starts, inspect the ground area to ensure it is capable of taking the weight of the plant and equipment and any applied load.
* Ensure all lifting operations are supervised by a competent person.
* Do not wrap chains and clings around the forks of a teleporter when used to lift loads.
* When using chains or slings with forks, use suitable fork clamps, with the chain or sling suspended from a suitable hook or shackle.
* When lifting with a teleporter, remove the forks and use a crane extension with hook or shackle.
* Do not carry out lifting operations in bad weather, or in the vicinity of overhead power lines.

* Ensure that the SWL is marked on all lifting equipment. Do not exceed the marked SWL under any circumstances. Fit alarms, where possible, to warn of over-loading.
* Ensure that all lifting lugs / lifting eyes have a marked SWL.
* Secure all ancillary equipment (for example, buckets used in connection with lifting equipment) at all times. Ensure that locking pins are in place.
* Ensure that all stillages used for lifting are in good condition and suitable for the task.
* Ensure all loads are secured and stable before lifting.
* When lifting pre-cast slabs or units, ensure that the lifting equipment is suitable for the task. Use special purpose lifting gear, as decided by a competent person. Ensure a Method Statement is in place for the lift.
* Attach block grabs correctly and use suitable netting to prevent loose blocks from falling.
* Ensure that skips being used as lifting gear are suitable for the task and are certified as lifting gear.
* Ensure that the area is cordoned off under lifting operations, and that barriers are used to prevent workers from entering the exclusion zone.

Applicable Legislation / Code of Practice

* Factories Act 1955, s.34 (1) (a).
* Chains, Ropes & Lifting Tackle (Register) Regulations, 1956.
* European Communities (Wire Ropes, Chains & Hooks) Regulations, 1979.
* Safety in Industry (Vehicle Lifting Tables & Other Lifting Machines) (Register of Examinations) Regulations, 1981.

* SHWW (General Application) Regulations, 1993, Part IV: Work Equipment Regulations; Part VI: Manual Handling Regulations.
* SHWW (Construction) Regulations, 2001.
* Work Equipment (Amendment) Regulations, 2001.
* SHWW Act, 2005.
* *Proposed SHWW (Construction) Regulations, 2006.*
* *Proposed SHWW (General Application) Regulations, 2006.*

* British Standards - BS 7121 Part 1-5.
* IS 360:2004. Code of Practice: *Safe Use of Cranes in the Construction Industry. Part 1: General.*

Required Documentation

* CR2 - Excavator or Loader Used as a Crane - Certificate of SWL.
* CR3 - Crane - Certificate of test & examination.
* CR3A - Crane - Report of anchoring / ballasting test.
* CR3B - Crane - Report of automatic safe load indicator test.
* CR4A - Lifting Appliances - Report of thorough examination (14 months / repair / first use)
* CR4B - Lifting Appliances - Weekly inspection report.
* CR6 - Chains, Slings, Rings, Links, Hooks, Plate Clamps, Shackles, Swivels & Eyebolts - Certificate of test & examination.
* CR6A - Chains, Ropes & Lifting Gear - Report of thorough examination.
* CR6B - Chains, Ropes & Lifting Gear - Report of annealing / heat treatment.

Training / Certification

* FÁS CSCS Telescopic Handler operation.
* FÁS CSCS Mobile Crane operation.
* FÁS CSCS Tower Crane operation.
* FÁS CSCS Slinger / Signaller.

Further Information

* ASCE: *Crane Safety on Construction Sites.*
* HSE: *Health & Safety in Construction*, HSG150(rev.).
* HSE: *Safe Working with Lift Trucks*, HHS(G)6.
* HSE: *Specification for Automatic Safe Load Indicators.*

Cranes & Lifting operations - see p. 84 / 136

LIGHTING

See also: Office Work / Welfare

Hazards → Risks

Inadequate lighting
Faulty earthing

Eye sight damage
Slips / trips / falls
Lacerations / abrasion
Sprains / strains
Electrocution

Controls → Managing the Risks

* Provide suitable /safe lighting.
* Provide access lighting, sufficient to carry out all operations (main contractor's responsibility).
* Provide task lighting, if required by the worker (sub-contractor's responsibility).
* Ensure that lighting is in place on all access routes for vehicles and pedestrians.
* Ensure that lighting is sufficient, especially where lifting operations are in progress.
* Ensure all areas with openings are adequately lighted and that warning signs are in place.
* Fit emergency lighting where special risks arise if the artificial light were to fail.
* Ensure that all lighting installed is checked regularly by a competent person.
* Ensure that all faulty lighting is marked and removed off-site until repaired by a competent person.

Applicable Legislation

* SHWW (General Application) Regulations, 1993, Parts IV, VII & VIII; Second, Third & Fourth Schedules.
* SHWW (Construction) Regulations, 2001, Part 4: Reg.17.
* SHWW Act, 2005.
* *Proposed SHWW (Construction) Regulations, 2006.*
* *Proposed SHWW (General Application) Regulations, 2006.*

Further Information

* Chartered Institution of Building Services Engineers Guides:
    o Code for Interior Lighting.
    o Lighting in Hostile & Hazardous Environments.
    o Emergency Lighting.
* HSE: *Lighting at Work*. HS/G38.

Lighting on a crane

NOTES

LONE WORKING

See also: Site Security

Who are lone workers?

Lone workers are those who work by themselves without close or direct supervision. They include:

* People who work separately from others.
* People working outside normal hours - for example, security, maintenance or repair staff.
* Mobile workers working away from their fixed base.

Lone workers are found in a wide range of situations, including construction, plant installation, maintenance, cleaning work, electrical repairs, painting and decorating, and vehicle recovery.

Hazards → Risks

Lone working
Accidents   **Injury**

Violence

Controls → Managing the Risks

* Assess risks to lone workers and take steps to avoid or control risk, where necessary (**employer's responsibility**).
* When Risk Assessment (see below) shows that it is not possible for the work to be done safely by a lone worker, make arrangements to provide help or back-up, including:
  o Supervisors periodically visiting and observing people working alone.
  o Regular contact between the lone worker and his / her supervisor, using telephone or radio.
  o Automatic warning devices that operate if specific signals are not received

---

Risk Assessment Checklist

> Does the workplace present a special risk to the lone worker?
> Is there a safe way in, and a way out, for one person?
> Can any temporary access equipment that is necessary, such as portable ladders or trestles, be safely handled by one person?
> Can all the plant, substances and goods involved in the work be safely handled by one person?
> Does the work involve lifting objects too large for one person, or is more than one person needed to operate essential controls for the safe running of equipment?
> Is there a risk of violence?
> Are women especially at risk, if they work alone?
> Are young workers especially at risk, if they work alone?
> Is the person medically fit and suitable to work alone?
> What happens if the person becomes ill, has an accident, or there is an emergency?

periodically from the lone worker - for example, systems for security staff.

o Other devices designed to raise the alarm in the event of an emergency, and which are operated manually or automatically by the absence of activity.

o Checks that a lone worker has returned to their base or home on completion of a task.

o Access to adequate First Aid facilities - ensure that mobile workers carry a First Aid kit suitable for treating minor injuries.

Applicable Legislation

★ SHWW (General Application) Regulations, 1993.

★ SHWW (Construction) Regulations, 2001.

★ SHWW Act, 2005.

★ *Proposed SHWW (Construction) Regulations, 2006.*

★ *Proposed SHWW (General Application) Regulations, 2006.*

Further Information

★ HSE: *Working Alone in Safety*.

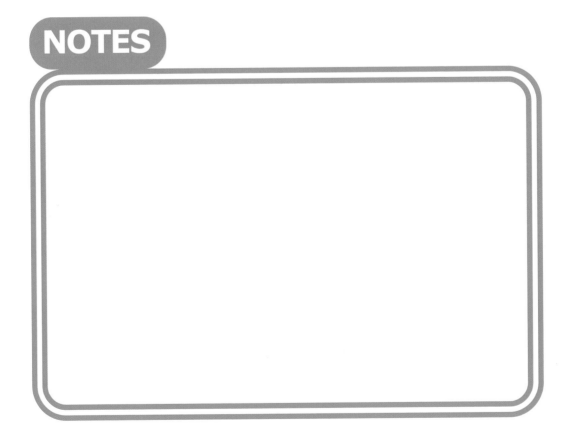

MACHINERY - GENERAL

See also: **Dumpers & Dump Trucks / Excavators / Lifting Equipment & Operations / Saws / Teleporters**

Hazards → Risks

Unsuitable / unsafe machinery

Operating machinery on uneven /
sloped surfaces

Untrained staff / unauthorised use /
carrying passengers

→ **Injury / crushing Death**

Hydraulics / operating attachments /
raised body parts / elevated equipment /
attachments

Pressure failure

Controls → Managing the Risks

* Before use, check all machines for suitability for the task - for example, SWL, boom length, type of attachments, buckets. etc.

* Under no circumstances use machines for work it is not designed for.

* Allow only authorised / competent workers to drive / operate machines.

* Allow only trained / competent banksmen (valid CSCS card) to give signals to machine drivers and always use them during any lifting operations.

* Fit all machines with roll-over protection / safety frame.

* Fit machines with suitable warning devices - for example, flashing beacons.

* Fit all machines with restricted visibility with additional devices such as CCTV, convex mirrors, etc.

* Remove the keys of the machine when it is not in use and park the machine so as not to present a danger to anyone.

* Carry out all operations as per the operator's manual of each machine. Keep a copy of the operator's manual in the cab of all machines.

* Wear seat belts.

Company vehicles come in all shapes and sizes

* Do not allow unauthorised riding on machines - one seat, one person. Do not allow riding on insecure positions, such as buffers or the running boards.
* Ensure that all plant and machinery adhere to site speed limits and that they conform to site traffic control plans.
* Protect all dangerous / moving parts with suitable guards.
* Do not bypassing or interfere with safety devices on the machines.
* Before adjusting / repairing / maintaining / leaving a machine, always turn off the engine.
* Under no circumstances rely on the hydraulics to maintain raised equipment while working underneath.
* Where required, support all raised machinery to prevent sudden collapse.
* Report all damaged / faulty machinery to the Site Manager / Safety Officer immediately and remove them from work areas.
* Secure all plant (within locked compound / fenced area) when left unattended, to prevent unauthorised use, especially at the end of the working day.
* Do not use mobile phones while operating any machinery.
* Do not operate machines while under the influence of alcohol / drugs, including prescribed drugs.
* Do not wear loose clothing, jewellery, belts, etc. while operating machinery.

Site traffic control includes separate car parks for contractors' vehicles and pedestrian crossings at approriate places

Worker / Public Safety

Always keep people at a safe distance from working plant and use barriers, where possible - whether they are workers on site or members of the public, the separation of

Scaffolding - see p. 186

pedestrians from all plant operations is important. Identify pedestrian routes clearly and erect suitable signage.

Site traffic control
Ensure that the construction site has a site traffic plan to control traffic movement, which ideally segregates construction machines from cars. Explain this plan at Induction training and display it in a prominent location - for example, Canteen. Plans may include the use of warning signs, bollards, stop / go systems, ramps, temporary traffic lights and flagsmen. For construction work close to roads or road works, liaise as necessary with local Gardaí.

Applicable Legislation

* SHWW (General Application) Regulations, 1993, Part IV: Work Equipment Regulations.
* SHWW (Construction) Regulations, 2001, Part 11: Transport, Earthmoving & Materials Handling Machinery, Regs.41-46.
* SHWW Act, 2005.
* *Proposed SHWW (Construction) Regulations, 2006.*
* *Proposed SHWW (General Application) Regulations, 2006.*

Training / Certification

* FÁS CSCS Telescopic Handler operation.
" FÁS CSCS Mobile Crane operation.
* FÁS CSCS Tower Crane operation.
* FÁS CSCS Slinger / Signaller.
* FÁS CSCS 360° Excavator operation.
* FÁS CSCS 180° Excavator operation.
* FÁS CSCS Site Dumper operation.
* FÁS CSCS Articulated dumper operation.

Further Information

* HSA: *Safety of Workplace Vehicles.*

MACHINERY - MOVING PARTS

See also: Cement Mixers / Chainsaws

Hazards → Risks

Clearing obstructions
Faulty / defective machinery
Unsupervised / untrained
  persons
Unauthorised use

**Entanglement /
  burns / lacerations
Trapped fingers /
  hands / hair
Traumatic amputation
Crush injuries
Death**

Controls → Managing the Risks

✴ Observe manufacturer's instructions on operation / maintenance and do not make modifications or make-shift repairs.

✴ Do not bypass or interfere with safety devices on machines.

✴ Allow only trained / authorised workers to service machines.

✴ Do not allow persons under 18 to service plant / machinery, unless under the supervision of a competent person.

✴ Ensure that all staff know the 'emergency stop' procedure for each machine.

✴ Report all defects in machinery immediately to site management and shut down and isolate the machine so as not to endanger anybody.

Applicable Legislation

✴ SHWW (General Application) Regulations, 1993, Part IV: Work Equipment Regulations.

✴ SHWW (Construction) Regulations, 2001.

✴ SHWW Act, 2005.

✴ *Proposed SHWW (Construction) Regulations, 2006.*

✴ *Proposed SHWW (General Application) Regulations, 2006, Part III.*

Training Requirements

✴ Abrasive Wheel training.

Further Information

✴ Manufacturer's instructions / suppliers' guidelines.

MACHINERY - REFUELLING

See also: Diesel Fumes / Fire

Hazards → Risks

Fire / explosion   →   **Burns**
                                 Scalds

Controls → Managing the Risks

* Make available suppliers' MSDS sheets for all fuels / chemicals, etc.
" Store fuels away from flammable materials / openings / sewers / excavations.
* Keep all areas clear of obstructions and substances that contribute to slips / trips / falls.
* Do not allow sources of heat to be present during refuelling.
* Ensure that suitable fire extinguishers are available in each machine.
* No smoking / use of mobile phones in refuelling area.
* Use only suitable / approved containers for storing / transporting fuels.
* Switch off all machinery before any refuelling / maintenance.
* Carry out refuelling in a well-ventilated area.
* If possible, perform all operations requiring the running / idling of the engine in a well ventilated area (ideally, outdoors) and for the minimum time.
* Before adjusting / repairing / leaving a machine, always turn off the engine.

Lubricants / greases:
* Avoid prolonged and repeated contact with mineral oils / used oil.
* Exercise good personal hygiene - wash hands with a good hand cleaner.
* Take extra care when handling used oils / greases containing lead.
* Wear appropriate PPE and wash protective clothing regularly.
* Do not use petrol / paraffin / solvents to clean oil / grease from the skin.

Applicable Legislation

* SHWW (General Application) Regulations, 1993, Part IV: Work Equipment Regulations.
* SHWW (Chemical Agents) Regulations, 2001.
* SHWW (Construction) Regulations, 2001, Part 9: Dangerous or Unhealthy Atmospheres, Regs.36-38.
* SHWW Act, 2005.
* *Proposed SHWW (Construction) Regulations, 2006.*
* *Proposed SHWW (General Application) Regulations, 2006.*

Manual Handling

See also: Blocks & Bricks

What is manual handling?

More than one-third of lost time accidents reported to HSA are the result of injuries sustained during manual handling operations. The most common injuries arising from manual handling are musculo-skeletal problems, meaning any strain to muscles, ligaments or tendons.

Hazards → Risks

Back injury   Stacking / loading /

/ hand injury /
delivering / sorting
supplies

→

hernia / muscle tear
Strains / sprains
Cramp / crush injuries

Controls → Managing the Risks

* Carry out task-specific manual handling Risk Assessments on all work activities.
* Use mechanical lifting devices, where possible - for example, teleporter / forklift / chains / slings, etc.
* Ensure that all workers at risk receive appropriate manual handling training and keep records of same.
* Organise the stores / storage area and adopt good house-keeping techniques.
* Avoid two-person lifts, if possible. Allow only suitably trained workers to carry out two-person lifts.
* Keep all areas clear of obstructions and substances that contribute to slips / trips / falls.
* Do not allow workers to carry anything that obscures their vision.
* Avoid direct handling of sharp-edged items. Remove or wrap sharp edges first.
* Always ensure nails are removed or hammered back on all scrap timber.
* Do not allow a worker who has a history of back trouble to undertake any manual handling task.
* Arrange work to avoid over-reaching or twisting when manual handling.
* Avoid tasks that require reaching over shoulder-height and / or twisting of the lower back region.
* Store heavy goods ideally between knuckle- and shoulder-height.

* Wear suitable PPE, such as safety footwear, protective gloves and overalls, when handling materials, and ensure that there is no risk of entanglement.
* Carry out periodic audits on manual handling techniques to identify any lapses in good lifting practices.

Applicable Legislation

* SHWW (General Application) Regulations, 1993, Part VI: Manual Handling of Loads, Reg.27 & 28; Eighth Schedule.
* SHWW (Construction) Regulations, 2001.
* SHWW Act, 2005.
* *Proposed SHWW (Construction) Regulations, 2006.*
* *Proposed SHWW (General Application) Regulations, 2006, Part V.*

Cement must be bought in 25kg bags, which can be lifted easily

Training / Certification

Manual handling training is a legal requirement, where it is identified that manual handling operations are required for work activities. This applies to all construction sites - due to the nature of construction work, materials are handled every day.

Manual handling training involves learning the correct lifting techniques, the principles of lifting and the anatomy of the back, which allows employees to adopt best practice. Records of training must be retained on file.

Further Information

* HSA: *Guidance on Training in the Manual Handling of Loads.*
* HSA: *Handle with Care - Safe Manual Handling.*
* HSE: *Backs for the Future, Safe Manual Handling in Construction*, HSG149.
* HSE: *Preventing Injuries from Manual Handling of Sharp Edges in the Engineering Industry*, EIS16.

MEWPs

See also: Fall Protection Equipment / Working at Height

What are MEWPs?

Mobile Elevating Working Platforms (MEWPs) are better known as 'cherry-pickers' and 'scissors lifts'.

Hazards → Risks

Working at heights - overhead
  power lines / falls /
  overturning / collision /   →
  unauthorised use
Collapse / over-turning
 Faulty outriggers
Unstable ground conditions

Persons falling
Equipment / materials
 falling
Injury - trapped against
 a fixed structure
Electrocution

Controls → Managing the Risks

✱ Allow only authorised / competent persons (over 18) to operate MEWPs.

✱ Maintain all MEWPs as per manufacturer's instructions.

✱ Ensure all maintenance is carried out by a competent person.

✱ Ensure all MEWPs have daily safety checks and a weekly inspection.

✱ Ensure the MEWP is suitable for the task (ground conditions, working height, the tasks, range / sensitivity of movement, anticipated load, etc).

✱ Do not operate MEWPs on unstable ground or soft ground conditions.

✱ Use mechanical stabilisers, where necessary. Ensure that stabilisers are extended before the platform is raised.

✱ Check work area for localised features - for example, manholes, service ducts, potholes, etc. (**Note:** A hole of as little as 75mm deep can cause an MEWP to over-turn). Use temporary covers, strong enough to withstand applied pressure, to cover localised features.

✱ Do not operate MEWPs near overhead power lines.

✱ Ensure that a safe system is in place to prevent people from being struck by the platform or the platform coming into contact with obstructions that may cause it to tip.

✱ Do not allow operators to climb out of, or over-reach, while working in the platform.

✱ Before use, ensure the safety rail is secured.

✱ Ensure that the doors on the basket are securely locked, before raising.

✱ Check that all the emergency stop devices are in good working order.

* Ensure that all operators wear suitable fall protection PPE, anchored within the platform, while working in an elevated platform.
* Ensure that all fall protection / work restraint systems for use on a MEWP are selected by a competent person. (A work restraint system for use on a MEWP should normally be a combination of a full body harness (BS / EN 361) and a lanyard (BS / EN 354) - not normally shock-absorbing.)
* Ensure that workers keep both their feet on the platform at all times.
* Do not move the machine with the platform in a raised position, unless designed to do so.
* Beware of obstacles protruding from roofs or sides.
* Keep the platform tidy and free of materials. Clear the basket completely at end of work.
* Do not use a MEWP as a crane.
* Secure MEWPs when left unattended, to prevent unauthorised use, especially at the end of the working day.

A cherry picker (top) and scissors lift (above)

Applicable Legislation / Standards

* SHWW (General Application) Regulations, 1993, Part IV: Work Equipment Regulations.
* SHWW (Construction) Regulations, 2001.
* SHWW Act, 2005.
* *Proposed SHWW (Construction) Regulations, 2006.*
* *Proposed SHWW (General Application) Regulations, 2006.*
* BS / EN 361 - full body harness.
* BS / EN 354 - lanyard.

Training / Certification

* Ensure a MEWP has a thorough inspection by a competent person at least once every six months.
* Inspections may be more frequent depending on the use and operating conditions, and intervals should be stated in the examination schedule.
* Normally, a MEWP has daily checks and weekly inspections.
* Operatives should receive training in safe use and operation of MEWPs.

Further Information

* HSE: *Preventing Falls from Boom-type Mobile Elevating Work Platforms*, MISC614.

MOBILE PHONES

Hazards → Risks

Loss of concentration
Distractions

 → **Accidents**

Controls → Managing the Risks

* Do not use mobile phones:
  o When operating plant & machinery.
  o When a vehicle / machine in is motion.
  o In areas where machines are refuelling.
* Keep the overall use of mobile phones on construction sites to a minimum. The preferred method of communication on-site is via hands-free two-way radio - especially for lifting operations.

The Road Traffic Regulations, 2002, state 'the driver of a mechanically-propelled vehicle shall not hold or have about their person a mobile phone or other similar apparatus while in the said vehicle, except when it is parked'.

> **Note: Hand-held CB2 radios, two-way private radio systems and walkie-talkies are also covered by the ban. The use of hands-free phones and equipment is permitted.**

Applicable Legislation

* SHWW (General Application) Regulations, 1993.
* SHWW (Construction) Regulations, 2001.
* Road Traffic (Construction, Equipment & Use of Vehicles) (Amendment) (No.2) Regulations, 2002.
* SHWW Act, 2005.

Hand-held CB2 radios, two-way private radio systems and walkie-talkies are banned.

Only hands-free phones and equipment are permitted on site.

NOISE

See also: PPE

Hazards → Risks

Inability to hear other sounds,
  instructions and warnings
Affects concentration
Fatigue

→

Proneness to accidents
Noise-induced hearing
loss / deafness

Controls → Managing the Risks

Three key control measures are:
* Access - noise survey.
* Eliminate - remove noise sources from site.
* Control - measures to prevent / reduce exposure.

General control measures include:
* Use physical noise barriers, where possible - by fitting of silencers, etc.
* Ensure that a noise survey is carried out on a regular basis by a competent person, and that a Risk Assessment is prepared.
* Put in place a control programme and create ear protection zones.
* Isolate plant & machinery that emits high levels of noise, where possible.
* Make sure all forms of ear protection are available to all workers.
* Provide training and information to workers on the dangers of noise and the use, care and maintenance of PPE.
* Wear hearing protection at all times when using or working in the vicinity of operating rock-breakers / scabblers / kango hammers / angle-grinders or any other equipment emitting high noise levels.

Reduce - or eliminate -
sources of excess noise,
such as rockbreakers
(above) and using a jack
hammer (below)

Ear protection

Disposable ear plugs	Correct insertion essential Not-re-usable Handle only with clean hands
Re-usable ear plugs	Need regular and careful washing Supplied and fitted by competent person Fitted for individual worker Dirt can cause ear irritation
Ear defenders (muffs)	Must be correct for job If loose fitting, damaged or worn - not effective Hair style or glasses may cause problems

Common Construction Noise Exposure Levels

Occupations	dB(A)	Machine-types	dB(A)
Site Manager (up to 50% day on site)	**<80**	Pneumatic Hammer	**103-113**
Bricklayer	**81-85**	Jack hammer	**102-111**
Carpenter	**86-96**	Consaw	**98-102**
Concrete Worker	**89**	Skilsaw	**98-102**
Engineer (supervising pour)	**96**	Bulldozer	**93-98**
Foreman (supervising workers)	**80**	Earth Tamper	**90-96**
Machine Driver: Front end loader	**86-94**		
Dumper	**85+**		
Excavator	**<85**		
Roller	**85+**		
Wheeled loader	**89**		

Applicable Legislation

* SHWW (General Application) Regulations, 1993, Sixth Schedule (5).
* Hearing Injury Act, 1998.
* SHWW (Construction) Regulations, 2001.
* Noise Regulations, 1990.
  o Action Level 1 - 85 dB (A) - you have to shout to be heard at 2m.
  o Action Level 2 - 90 dB (A) - you have to shout to be heard at 1m.
  o Action Level 3 - 200 pascals (A) - you cannot be heard talking beside someone.
* SHWW Act, 2005.
* *Proposed SHWW (Construction) Regulations, 2006.*
* *Proposed SHWW (General Application) Regulations, 2006.*

> **Note:** **By February 2006, all EU Member States are obliged to implement the 2003 Directive imposing new safety standards for noise at work, which include an increase in the action levels. New Noise Regulations are expected in 2006.**

Further Information

* Construction Industry Research & Information Association: *A Guide to Reducing the Exposure of Construction Workers to Noise*, Report 120.
* HSA: *Guidelines to the Noise Regulations*.
* HSA: *Is Your Work Making You Deaf?*
* HSE: *Ear Protection*, INDG 298.
* HSE: *Noise in Construction*, INDG 127.

NOTES

NON-NATIONAL WORKERS

'Over the last three years, 4,157 work permits have been issued to non-nationals to work in the construction sector. In 2004 alone, 1,213 permits were issued to nationals from 39 separate countries. Over 60% were issued to the nationals of three countries - Turkey (411), Poland (255) and Romania (120). Non-nationals make up less than 4% of those employed in this sector but present us with real language issues when it comes to ensuring safety on site.' (HSA, Spring Newsletter, 2005)

Hazards → Risks

Communication - unable to   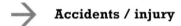 **Accidents / injury**
understand signs / warnings

Controls → Managing the Risks

* Ensure all new employees are in possession of a current Safe Pass card.
* Ensure all non-nationals receive Induction training, before working on-site.
* Ensure all non-nationals have sufficient competency of the English language so as to carry out their work and understand instructions / training given.
* Ensure that all workers with poor English are under the supervision of at least one worker who is competent in English.
* Use the HSA 'Safe System of Work Plan' for hazard identification and controls on sites where there are large numbers of non-national workers.
* Ensure that site signage includes images, where possible, to warn of hazards.

Applicable Legislation

* SHWW (General Application) Regulations, 1993.
* SHWW (Construction) Regulations, 2001.
* Equality Act, 2004.
* SHWW Act, 2005.
* *Proposed SHWW (Construction) Regulations, 2006.*
* *Proposed SHWW (General Application) Regulations, 2006.*

Further Information

* CIF has developed *Safety Handbooks*, translated into foreign languages to accommodate the growing numbers of non-national workers on Irish sites. For further information, contact CIF.
* HSA: *Safe System of Work Plan: Ground Works / House-building* (also available in Polish and Turkish editions).

OFFICE WORK

**See also: House-keeping / Lighting / Visual Display
Units**

Office work is often the forgotten hazard on construction sites. Although the risk of injury is low in comparison to construction activities, office workers are still susceptible to hazards.

Hazards → Risks

Open drawers / cabinets	**Lacerations / crushing**
Trailing leads	**Slips / trips / falls**
Slippery surfaces / poor	**Back / hand / head injuries**
lighting / poor house-keeping	

Controls → Managing the Risks

* Put in place a good house-keeping / cleaning programme to manage the orderly movement / cleaning of all areas and materials / persons from the point of entry to exit and within the site.
* Ensure that all staff practice good house-keeping and a 'clean as you go' policy.
* Keep walkways / aisles free of obstructions.
* Store bags / briefcases in lockers / presses.
* Put away sharp objects and keep desk / filing cabinet drawers closed when not in use.
* Report immediately to management any defects in office furniture.
* Maintain lighting at a level to enable staff to carry out their work safely and to permit safe passage.

Applicable Legislation

* SHWW (General Application) Regulations, 1993, Part IV: Work Equipment Regulations.
* SHWW (Construction) Regulations, 2001.
* SHWW Act, 2005.
* *Proposed SHWW (Construction) Regulations, 2006.*
* *Proposed SHWW (General Application) Regulations, 2006.*

Further Information

* HSA: *Guidelines to the Health & Safety of Office Workers.*
* HSE: *Officewise.*

OPENINGS

See also: Working at Height

Hazards → Risks

Falls   →   **Injuries**
Falling objects   **Death**

Controls → Managing the Risks

* Protect openings greater than 2 metres with guard-rails or cover them with materials strong enough to take the intended use - for example, sheet metal.
* Ensure the covering is capable of supporting the weight of the materials / staff placed upon it and never over-load it.
* Construct all coverings to withstand impact, and ensure that they are sufficiently and securely fixed in position to prevent accidental dislodgement.
* Mark all coverings with a warning sign - for example, 'Hole below - do not remove'.
* Where the protective guard-rails / barriers / covering are removed, do not leave the opening unattended. Replace them as soon as is practicable.
* Carry out regular checks to ensure that the openings are safe and that protective measures are not tampered with.

Use tape, bollards, hazard lights or fencing (below) to screen openings and make them safe

Applicable Legislation

* SHWW (General Application) Regulations, 1993.
* SHWW (Construction) Regulations, 2001, Part 13: Working at Heights, Regs.74 & 75.
* SHWW Act, 2005.
* *Proposed SHWW (Construction) Regulations, 2006.*
* *Proposed SHWW (General Application) Regulations, 2006.*

Working at height - see p.224

PPE

See also Dust / Lead / Noise / RPE

Hazards → Risks

Inadequate or inappropriate PPE   →   **Injuries**
                                          Death

Controls → Managing the Risks

General

* Ensure all PPE (helmets / safety footwear / high visibility vests, etc.) is selected by a competent person based on Risk Assessments of the work to be carried out.
* Ensure all PPE meets the minimum Irish, British or European standards.
* Ensure all workers are trained in the correct use / storage / care and maintenance / inspection of PPE.
* Ensure that all workers wear PPE provided at all times.
* Report damage / faults / defects in PPE to site management immediately.

Safety Helmets

Do:

* Wear the helmet the right way around - it does not give proper protection when worn back to front.
* Keep a supply of helmets for visitors on site.
* Wear a chin strap or tighten screw-backs if bending forward or down, or working at height.

Don't:

* Use helmet as a handy basket - it is designed to fit on your head, not for carrying nails!!
* Store in heat or direct sunlight. Excessive heat and sunlight can quickly weaken the plastic.
* Modify, cut or drill helmet.
* Share your helmet with other workers on site.

Safety Boots

* Required on all building sites. Do not give access to any person not wearing safety boots.
* Ensure that safety boots have steel toecaps and sole protection (puncture-proof soles).

* Waterproof boots are also advisable. Muddy or wet conditions will require safety rubber boots.

Hi-visibility Vests
* Compulsory on Irish construction sites. There are many versions of vests / jackets / pants, etc.
* Carry out a Risk Assessment to identify the level of hi-visibility clothing required by workers.

Hand Protection - Gloves
* Must be used when handling chemicals, abrasives, sharp or excessively coarse material, cement or concrete.
* Must be suitable for the job. If in doubt, seek advice from the Safety Officer.
* Latex gloves must not be reused - always dispose of when the job is completed.

Body Protection
* Overalls - required for work with chemicals / maintenance / repairs.
* Weatherproof clothing - required for outdoor work.

Eye Protection / Face Protection
* Includes spectacles / goggles / face shields, etc.
* Always wear eye / face protection when grinding, welding, burning, using chemicals and most power tools.
* Wear arc-welding masks and helmets when welding.

Ear Protection
* Ensure that noise levels are tested and monitored by a competent person.
* Ensure that appropriate ear protection is selected, based on the results of a Risk Assessment.
* Train all workers on how to use ear protectors correctly.
* Disposable ear-plugs:
  o Correct insertion essential
  o Not-re-usable - discard after use.
  o Handle only with clean hands.
  o Do not share.

* Re-usable ear-plugs:
  o Need regular and careful washing.
  o Supplied and fitted by competent person.
  o Fitted for individual worker - do not share.
  o Dirt can cause ear irritation.

* Ear-defenders (muffs):
  o Must be correct for job.
  o If loose fitting, damaged or worn - not effective.
  o Hair style or glasses may cause problems.
  o Stored in suitable containers.

Respiratory Protection

* Ensure that all RPE is selected by a competent person.
* Ensure that training is provided to all workers before use and that all equipment is tested before use.
* Adhere to manufacturer's instructions in use of all RPE.
* Store and maintain equipment in good working order.
* Report all defects immediately to site management.

Safety Harnesses

* Ensure all harnesses are selected by a competent person.
* Fit harnesses to the individual worker's needs - based on height / weight / work to be carried out.
* Train all workers in the correct use of safety harnesses / safe storage / inspection.
* Train all workers in emergency procedures and rescue techniques.
* Ensure that all work requiring the use of safety harnesses is under the supervision of a competent person.
* Ensure all harnesses are inspected by a competent person on a regular basis.

Applicable Legislation / Standards

* Abrasive Wheels Regulations, 1982.
* Asbestos Regulations, 1989 to 2000.
* Noise Regulations, 1990.
* SHWW (General Application) Regulations, 1993, Part V: Personal Protective Equipment.
* Biological Agents Regulations, 1994 to 1998.
* Chemical Agents Regulations, 1994.
* Carcinogens Regulations, 2000.
* Ionising Radiation Order, 2000.
* SHWW (Construction) Regulations, 2001.
* SHWW Act, 2005.
* *Proposed SHWW (Construction) Regulations, 2006.*
* *Proposed SHWW (General Application) Regulations, 2006, Part IV: PPE; Schedule 3.*

* **Ear protection:** CE EN 352-2.
* **Eye / face protection:** BS EN 166.
* **Gloves:** BS EN 420.
* **Mechanical Hazards:** EN 388.
* **Chemical Hazards:** EN 374-3.
* **Hi-visibility vests:** IS EN 471:1994 Class 2 / 3.
* **RPE: BS 4275:** 1997 'Best Practice'.
* **Safety boots:** CE EN 345 / CE EN346.
* **Safety helmets:** EN 397.
* IS / EN 340:1994. *Protective Clothing - General Requirements*.

Further Information

* HSA: *Guidelines to The SHWW Act,1989 & The SHWW (General Application) Regulations, 1993: Part V: Provision of PPE* (pp.78-96).
* HSE: *PPE: Safety Helmets*, CIS50.

NOTES

Pre-Cast Elements & Components

See also: Concrete & Cement / Lifting Equipment & Operations / Steel Fixing / Working at Height

Hazards → Risks

Arrival and erection of
  pre-cast units on site
Over-turning, moving
  machinery, electricity,
  striking objects
Falls / falling objects

Injuries
Crushing
Death

Controls → Managing the Risks

✶ Ensure that the following pre-cast health and safety documentation is available on site before arrival of any pre-cast units:
  o Pre-cast company Safety Statement.
  o Site-specific Method Statement.
  o Responsibility check-list.
  o MSDS for pre-cast products.

✶ The control and monitoring of all pre-cast erection is the responsibility of the site manager and the sub-contractor.

✶ Allow only trained banksmen / slingers (valid CSCS cards) to direct unloading and to keep the area cleared of pedestrians and vehicles.

✶ Ensure that framework, walls, etc. is of adequate strength to receive pre-cast units.

✶ Ensure that all bearings (masonry, concrete, steel) are completed, fully cured and / or properly secured.

✶ Ensure that the pre-cast erection crew all wear suitable PPE.

✶ Ensure that all members of erection crews are trained in working at heights and the safe use and checking of safety harness.

✶ Do not carry out any work on floors below the erection level during erection of units.

✶ Ensure that all pre-cast floor slabs, etc are laid in a systematic pattern that allows the use of collective fall protection systems.

✶ Ensure that slab anchors are designed and planned to suit the specific job.

* Where propping is required, detail the location of prop points on the pre-cast drawing.

* Ensure that any alterations (cutting, notching, etc.) are approved by the pre-cast company before cutting.

* Stop all erection activities during high winds.

* Remove all casted slab eyes when pre-cast is installed.

* If Transport Anchor Systems are used to lift pre-cast

concrete units, ensure that these are controlled by a competent person at all times. The person in control of the operation is responsible for the selection of the ring clutches for the anchors.

Applicable Legislation

* SHWW (General Application) Regulations, 1993, Part IV: Work Equipment Regulations.

* SHWW (Construction) Regulations, 2001.

* SHWW Act, 2005.

* *Proposed SHWW (Construction) Regulations, 2006.*

* *Proposed SHWW (General Application) Regulations, 2006.*

NOTES

PREGNANT EMPLOYEES

See also: Lead / Welfare Facilities

The law in relation to pregnant employees applies to three categories of employees:

* A pregnant employee.
* An employee who has recently given birth.
* An employee who is breastfeeding.

Hazards → Risks

Collision
Fatigue
Stress
Physical limitations

→ **Back injury**
Miscarriage

Controls → Managing the Risks

* Pregnant employees must inform their employer as soon as practicable after they become aware of their condition, with the appropriate medical certification.
* Ensure that the Safety officer carries out a written Risk Assessment of all chemical / physical agents, processes and working conditions that may be a hazard to the pregnant employee.
* Following the Risk Assessment, take the necessary protective and preventative measures to safeguard the pregnant employee.
* High risk work activities include:
  o Biological agents - viruses, bacteria, etc
  o Chemical agents - carcinogens, mercury, lead, carbon monoxide.
  o Extremes of cold or heat.
  o Ionizing / non-ionizing radiation.
  o Manual handling.
  o Night work.
  o Noise.
  o Physical agents - shocks, vibration.
  o Working in compressed air.
* Adjust temporarily the working conditions / working hours of the pregnant employee, based on the results of the Risk Assessment.
* Assign alternative work to the employee, if the hazard cannot technically or feasibly be reduced.

* Provide rest areas for pregnant or nursing employees.
* Unless a complete Risk Assessment is carried out by a competent person, which indicates there will be no ill-effects to mother or baby, do not allow pregnant employees or breast-feeding mothers to work with:
  o Lead / lead substances.
  o Pressurised enclosures.
  o Toxoplasma.
  o Underground mine work.

Applicable Legislation

* SHWW (General Application) Regulations, 1993, Part III.
* Maternity Protection Act, 1994.
* SHWW (Night Work & Shift Work) Regulations, 2000.
* SHWW (Pregnant Employees, etc) Regulations, 2000.
* SHWW (Construction) Regulations, 2001.
* SHWW Act, 2005.
* *Proposed SHWW (Construction) Regulations, 2006.*
* *Proposed SHWW (General Application) Regulations, 2006, Part XI: Protection of Pregnant, Post-Natal or Breastfeeding Employees.*

> Note: **The proposed SHWW (General Application) Regulations 2006, Part XI, is intended to replace the SHWW (Pregnant Employees) Regulations, 2000, and Part XII: Night Work & Shift Work is intended to replace the SHWW (Night Work & Shift Work) Regulations, 2000.**

Further Information

* HSA: *Health & Safety at Work when Pregnant.*
* Chemicals classified as carcinogens / mutagens / toxic to reproduction: http://ecb.jrc.it/classification-labelling.

PUBLIC SAFETY

See also   **Falling Objects**

Hazards → Risks

Pedestrians / Vehicles -
access to premises -
moving transport
Unsecured materials
Collision

Crushing
Injuries
Death

Controls → Managing the Risks

* Ensure that access to the construction site is to authorised persons only (Safe Pass card holders). Use security personnel or a responsible person to control access.

* Ensure that all visitors to the construction site sign-in and are accompanied on their visit around the site.

* Ensure that the construction site is suitably fenced with barriers / hoarding, etc to separate all construction activities from members of the public.

* Especially on street-side works, ensure that suitably-designed hoarding is erected by a competent person to protect the public.

* Make safety arrangements to ensure that normal pedestrian and public vehicular traffic is not put at undue risk as a result or construction work.

* Erect suitable warning signs to forewarn of the known dangers at the entry gates to, and the boundary of, the construction site.

* Where members of the public must have access close to construction work, provide suitable and safe routes to protect them from construction activities. Also give consideration to persons with disabilities.

* Identify and mark suitably pedestrian crossings from parking areas on site.

* Devise and implement a site traffic plan, including speed limits posted with appropriate signage.

* Keep all areas along traffic routes clear of obstructions, plant & machinery, materials, etc.

* Protect all open or partially back-filled excavations / manholes and prevent access by suitable barriers and warning signs.

∗ Adhere to good house-keeping practices at all times. Keep all public areas clear of construction-related debris such as muck, dust, trip hazards, sharp objects, falling objects, etc.

∗ Remove bottom ladders on scaffolds and lock away all dangerous materials at night.

∗ Ensure that the construction site is secured at night by the Site Manager / security personnel.

∗ Check the site entrance and public roadway regularly for debris and remove. Put in place an effective house-keeping policy.

∗ Use safety netting / sheeting to protect the public at risk from falling objects.

Material deliveries:

∗ Under no circumstances attempt unloading while the driver is in the vehicle.

∗ Use mechanical means, where possible, for the unloading of materials - for example, a teleporter.

∗ If necessary, use a trained banksman (valid CSCS card) to guide reversing vehicles.

∗ Restrict access to the work area / goods inwards to authorised personnel only and display suitable signage.

∗ Secure all materials stored externally, so as to prevent toppling / rolling, etc.

∗ Control all entry / exits in hazardous conditions.

Applicable Legislation

∗ SHWW (General Application) Regulations, 1993.

∗ Occupiers' Liability Act, 1995.

∗ SHWW (Construction) Regulations, 2001.

∗ SHWW Act, 2005.

∗ *Proposed SHWW (Construction) Regulations, 2006.*

∗ *Proposed SHWW (General Application) Regulations, 2006.*

Use warning signs (top), mirrors (above) and hazard flashers (below) to ensure public safety

Road Works

See also: **Electricity / Excavations / Hand & Power Tools / Openings / Public Safety / Safety Signs**

Hazards → Risks

Inappropriate / poorly
  located / dirty / sullied signs
Faulty / inoperable traffic lights   →   **Collision**
Slippery / uneven surfaces /
  poor visibility
Debris on public roadway

Openings / excavations   →   **Falls / injury / collision**
Work over water   →   **Falls / back, hand or head injuries / drowning**

Underground / Overground
  cables / utilities   →   **Electrocution / burns**

Controls → Managing the Risks

* Before each section / sequence of work is undertaken, carry out a Risk Assessment to identify the most appropriate traffic layout.

* On the basis of the Risk Assessment, decide on the type / number of signs, cones, barriers, lights and any other control devices required.

* Place suitable warning signs (for example, 'Major Road Works Ahead' and 'Man-with-Shovel') at an appropriate distance (considering the topography / prevailing traffic speed on the road) from each end of the area of operations to forewarn all road users.

* Adopt a suitable daily cleaning programme to ensure that all signs / cones / drums are kept clear / clean.

* Ensure all signs / cones / drums are fitted with reflective bands.

* Replace all damaged / faulty signs / cones / drums immediately.

* Remove / cover signs, etc during all periods where no work is being undertaken and the signs are not relevant.

* Use a one-way shuttle operation, controlled by appropriate traffic light signals, as per Regulations, located at each end of the lead-in taper / safety zone (50 to 120 meters out, depending on road conditions / topography).

* Check / clean all lights daily.

* In the event of traffic light failure, control the entrance to the work area by persons equipped with 'Stop / Go' signs.
* Ensure that the size / width of the operational traffic lane is such that it can only accommodate one line of traffic at a time.
* Use brightly coloured tape only to mark out areas - do not use such tape as a fall protection barrier around an open trench / excavation.
* Backfill all road openings each evening or at the end of the work period.
* Where openings must remain, cover them with material capable of supporting the weight of the materials / staff placed upon them without overloading.
* Mark all coverings with a warning sign (for example, 'Hole Below - Do Not Remove') and ensure that they are suitably illuminated with warning lights / signs during lighting-up time.
* Where the protective guard rails / barriers / covering must be removed for temporary access / movement of materials, etc, do not leave the opening unattended. Replace the protective barriers as soon as is practicable.
* Make regular checks to ensure that the openings are safe and that protective measures are not tampered with.
* Check the public roadway immediately adjoining the site regularly for clay / stones / debris from vehicles / plant operating to and from the site. Adopt an effective cleaning programme, where necessary.
* Ensure that all personnel working on / close to the road wear high visibility clothing.

Electricity
* Always assume that underground cables are present until, confirmed otherwise.
* Consult ESB / safety file / appropriate authority before any work commences for the location / depth / voltage of cables.
* Locate and mark location(s) of all cables / utilities.
* Where doubt exists as to the exact location of a cable / utility, locate it by hand-digging.
* Do not use electrical tools / machinery closer than 0.5 metres to a known cable location.
* Ensure that all post / poles / boxes / chambers are properly supported / underpinned, where necessary.
* Erect suitable barriers / notices to ensure that vehicles are kept away, if possible, from overhead cables - otherwise, provide suitable warnings and suspended protections ('goal posts').
* Do not allow tipping operations under cables.

Work over water

* Before undertaking this work, carry out a Risk Assessment to identify the most appropriate / safest way to do so.
* Do not undertake this work, if the height of the water is such that it poses a threat to the safety of the workers involved.
* Design this work so that the workers do not have to enter the water.

Vehicles / plant and machinery

* Operate and maintain all vehicles / plant as per manufacturers' instructions.
* Adopt safe operating / driving practices at all times.
* Ensure all drivers / visitors receive instruction on the site traffic plan.
* Do not allow persons under 18 to drive/ operate vehicles / plant / machinery.
* Keep all areas clear of anything that will obstruct a driver's vision.
* Ensure high visibility vests are worn by staff.
* Take care when transferring goods from a trailer / truck / van / machinery to the stores / site.
* Under no circumstances allow loading / unloading to be attempted while the driver occupies the tractor unit / truck / van.
* Post signs alerting members of the public and workers to any known / foreseeable danger in good time.
* Ensure that all vehicle egress from the site is controlled by a security person / signaller / banksman.

Applicable Legislation

* SHWW (General Applications) Regulations, 1993.
* Road Traffic (Signs) Regulations, 1997.
* SHWW (Construction) Regulations, 2001.
* *Proposed SHWW (Construction) Regulations, 2006.*
* *Proposed SHWW (General Application) Regulations, 2006.*

Further Information

* ESB.
* Department of the Environment & Local Government:
  o *Guidelines for the Opening, Backfilling & Reinstatement of Trenches in Public Roads* (2002), DEHLG CH.
  o *Specification for Road Works*, NRA Publications Office.

Excavations & Excavators - see p. 100 / 103

ROOF WORK

**There is
NO safe
height**

See also Fall Protection Equipment / Working at Height

The best way to prevent a fall from a roof is not to go on the roof in the first place! So before starting any roof work, ask:

* Is the job necessary?
* Must it be done this way?
* If the work must be done, can the amount of time on the roof be reduced?
* Can roof sections be partially assembled at ground level?

Hazards → Risks

Falls from / through roof
Collapse of roof
Equipment / materials falling
   from / through roof

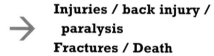

→   **Injuries / back injury /
paralysis
Fractures / Death**

Controls → Managing the Risks

* Allow only trained / competent workers to carry out roof works.
* Use suitable / approved / appropriate scaffolding.
* Erect guard-rails, toe-boards or suitable barriers at the edge or eaves level of the roof to prevent workers / materials falling.
* Inspect roofs, before starting work, to establish they are safe for the intended task, especially in cold / wet weather.
* Wear appropriate puncture-proof / non-slip footwear.
* Provide safe and suitable access / egress facilities.
* Where guard-rails / barriers cannot be provided, use a suitable safety harness, securely fixed.
* Carry out regular checks to ensure that the openings are safe and that protective measures are not tampered with.
* Do not allow workers to pass across, or work on, or from, fragile materials incapable of supporting their weight.
* Sign all such fragile roofs at the approach with a warning notice 'Danger - Fragile Roof'.
* Use roof ladders / crawling boards on sloping and fragile roofs.

* Do not pour any liquids or debris into roof drains.
* Cover all openings in roofs and mark them clearly.
* Ensure that the Safety Officer provides training to all workers on wearing / inspecting PPE.

Roof type	Hazards	Controls
Flat	Falls from edge of complete roof, edge where work is being carried out, openings, gaps, fragile roof lights, etc.	On roofs of less than 10° pitch, put guard-rails and toe-boards in place. Erect a barrier, set back from the edge. Wear appropriate fall protection PPE.
Pitched	Falls from eaves, slipping down roof, falling through the roof internally, from gable ends, etc.	Use roof ladders. 30° pitch - use crawling-boards / ladders / catch barriers. 50° + pitch - use a working platform. The longer the slope, the steeper the pitch, the stronger the edge protection required.

Falling materials can kill - therefore:

* Enclose rubbish chutes.
* Use a suitable method of lowering material to ground.
* Do not let material accumulate.
* Establish exclusion zones underneath, and adjacent to, roof work.
* Use debris netting, covered walkways, etc.
* Avoid large and heavy objects on roof.
* Ensure correct storage of all material.

Applicable Legislation

* SHWW (General Application) Regulations, 1993, Part IV: Work Equipment.
* SHWW (Construction) Regulations, 2001, Part 13: Working at Heights, Regs.76 & 77.
* SHWW Act, 2005.
* *Proposed SHWW (Construction) Regulations, 2006.*
* *Proposed SHWW (General Application) Regulations, 2006, Part XV: Working at Height.*
* *Code of Practice for Safety in Roofwork* (2005).

Further Information

* http://www.roofingcontractor.com.
* NISO: *Safety in Roofwork*, Construction Summary Sheet.

RPE

See also: PPE

What is RPE?

RPE stands for 'respiratory protective equipment'. Respiratory safety hazards are among the most difficult to protect against. The range of contaminants, working environments and number of further variables makes selecting respiratory equipment a potential 'minefield'. RPE is divided into two broad categories: Filter respirators and air supplied equipment.

> Note: **RPE should only be selected as a last resort. All other measures to control the hazard should be considered first - for example, substitution, elimination, separation, and engineering controls.**

Hazards → Risks

Inadequate or inappropriate RPE
Injuries
Suffocation
Death

Controls → Managing the Risks

* Where it is reasonably practicable to do so (employer's responsibility):
   o Enclose the work area and keep it under negative pressure.
   o Use controlled wet removal methods (for example, water injection / damping down the surface to be worked on).
   o Use a wrap-and-cut method or glove bag technique, where appropriate.
* Ensure that all RPE is selected by a competent person based on Risk Assessments of the work to be carried out.
* Ensure that all RPE provided is marked with a 'CE' symbol, which means it meets the minimum legal requirements, usually by conforming to a European Standard. The minimum required standard is BS 4275:1997, 'Best Practice'. The most significant impact of the standard is the introduction of Assigned Protection Factors (APFs), which are calculated in actual workplaces and give a more realistic indicator of a respirator's level of performance than previous methods.
* Ensure that all workers are trained in the correct use / storage / care and maintenance / inspection of all RPE.
* Ensure all workers wear RPE provided at all times.
* Adhere to manufacturer's instructions when using all RPE.
* Report damage / faults / defects to site management immediately.

Selection of RPE

RPE must be matched to:

* The exposure concentrations (expected or measured).
* The job.
* The wearer.
* Factors related to the working environment.

HSE has developed a RPE Selector Tool, designed to assist in the selection of RPE.

Suitable RPE means:

* It provides adequate protection (for example, reduces the wearer's exposure to asbestos fibres as low as is reasonably practicable, and in any case to below the control limits) during the job in hand and in the specified working environment (for example, confined spaces).
* It provides clean air and the appropriate flow rate during the wear period.
* The face-piece fits the wearer correctly.
* It is properly maintained.
* The chosen equipment does not introduce additional hazards that may put the wearer's health and safety at risk.

Also consider:

* The temperatures at which people will be working.
* The facial characteristics of the wearers (for example, beards, sideburns, stubble growth, glasses, etc).
* The medical fitness of the people who will wear the equipment.
* The length of time the person will have to wear the equipment.
* How comfortable it is, and whether people will wear it correctly for the required length of time.
* Whether the job involves extensive movements, restrictions and / or obstructions that need to be overcome while doing the job.
* The need to communicate verbally during work.
* The effects of other personal protective equipment and other accessories on RPE (for example, unmatched goggles may affect the face seal provided by the face-mask; jewellery may interfere with the performance of the RPE).

Note: People come in all sorts of shapes and sizes. Therefore, one particular size or type of RPE is unlikely to suit everyone. In addition, the performance of face-pieces (filtering face-pieces, half- and full-face masks) depends on achieving a good contact between the wearer's skin and the face seal of the mask. To make sure that the selected face-piece can provide adequate protection for the wearer, the initial selection should include fit testing. For those wearing glasses, full-face masks that allow the fixing of special frames inside the mask may be considered.

Applicable Legislation / Standards

* SHWW (General Application) Regulations, 1993, Part V: Personal Protective Equipment.
* SHWW (Construction) Regulations, 2001.
* SHWW Act, 2005.
* *Proposed SHWW (Construction) Regulations, 2006.*
* *Proposed SHWW (General Application) Regulations, 2006.*

* BS 4275:1997. 'Best Practice'.

Further Information

* HSE: *Respiratory Protective Equipment at Work: A Practical Guide* (rev.) (includes RPE Selector tool).
* HSE: *Selection of RPE Suitable for Use with Wood Dust*, WIS14.
* HSE: *Selection of Suitable RPE for Work with Asbestos*, HSG53.

NOTES

Lifting Equipment & Operations - see p. 136

SAFETY SIGNS

Hazards → Risks

Unknown dangers
Confusion / lack of   →   **Accident / injury**
  communication

Controls → Managing the Risks

✱ Prominently display pictorial safety / warning signs showing Access / Prohibition, Fire Exits, Fire Points, Flammable substances, PPE required, Dangerous Materials, Traffic & Pedestrian control.

✱ Report all damage to, and theft of, safety signs to the Safety Officer immediately.

Shape					
Colour	**Meaning / Purpose**	**Examples of Use**			
Red	Stop Prohibition	Stop signs Emergency shut-down signs Prohibition signs	**Prohibition**		**Fire-fighting equipment**
Yellow	Caution Possible danger	Identification of dangers (fire, explosion, radiation, chemical hazards, etc.) Identification of steps, dangerous passages and obstacles	Caution Possible danger		
Green	No danger First Aid	Identification of emergency routes and exits Safety showers First Aid stations and rescue points			**No danger Rescue equipment**
Blue	Mandatory signs Information	Obligation to wear individual safety equipment Location of telephone	**Mandatory**		**Information Instruction**

Adapted from HSA: *Obligatory Safety Signs*.

Applicable Legislation

*   SHWW (General Application) Regulations, 1993.
*   Signs Regulations, 1995.
*   SHWW (Construction) Regulations, 2001.
*   SHWW Act, 2005.
*   *Proposed SHWW (Construction) Regulations, 2006.*
*   *Proposed SHWW (General Application) Regulations, 2006, Part XIII: Safety Signs at Work.*

> Note: **The proposed SHWW (General Application) Regulations, 2006, Part XIII, are intended to replace the Signs Regulations, 1995.**

Further Information

*   HSA: *Obligatory Safety Signs.*

SAWS

See also: Hand & Power Tools / Machinery - General

Hazards → Risks

Incorrect adjustments of guards 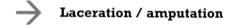   **Laceration / amputation**
Inexperienced operators

Controls → Managing the Risks

* Allow only trained / competent workers to use saws unsupervised.
* Adhere to manufacturers' instructions at all times.
* Ensure that machine-guards are in position and maintained correctly.
* Maintain blades at the minimum height and in first class condition.
* Always keep hands clear of the blade and never closer than 500 mm.
* Wear appropriate PPE.

Before starting to cut with powered hand-saws:
* Wear safety glasses or a face-shield.
* Make sure guards, if present, are installed and are working properly.
* Position the saw beside the material before cutting, and avoid entering the cut with a moving blade.

Working safely with powered hand-saws:
* Disconnect the power supply before changing or adjusting blades.
* Use lubricants when cutting metals.
* Keep all cords clear of cutting area.
* Remember that sabre-saws cut on the up-stroke.
* Secure and support stock as close as possible to the cutting line to avoid vibration.
* Keep the base or shoe of the saw in firm contact with the stock being cut.
* Select the correct blade for the material being cut and allow it to cut steadily. Do not force it. Clean and sharp blades operate best.
* Set the blade to go no further than 0.32 to 0.64 cm (1/8 to 1/4 inch) deeper than the material being cut.
* Do not start cutting until the saw reaches its full power.
* Do not force a saw along or around a curve. Allow the machine to turn with ease.
* Do not insert a blade into, or withdraw a blade from, a cut / lead hole while the blade is moving.
* Do not put down a saw until the motor has stopped.
* Do not reach under or around the stock being cut.

* Maintain control of the saw always. Avoid cutting above shoulder-height.

Starting an external cut:

* Place the front of the shoe on the stock.
* Make sure that the blade is not in contact with the material or the saw will stall when the motor starts.
* Hold the saw firmly down against the material and switch the saw on.
* Feed the blade slowly into the stock, maintaining an even forward pressure.

Starting an inside cut:

* Drill a lead hole slightly larger than the saw blade. With the saw switched off, insert the blade in the hole, until the shoe rests firmly on the stock.
* Do not let the blade touch the stock until the saw has been switched on.

Applicable Legislation

* SHWW (General Application) Regulations, 1993, Part IV: Work Equipment Regulations.
* SHWW (Construction) Regulations, 2001.
* SHWW Act, 2005.
* *Proposed SHWW (Construction) Regulations, 2006.*
* *Proposed SHWW (General Application) Regulations, 2006.*

SCAFFOLDING

See also: Blocks & Bricks / Fall Protection Systems / Falling Objects / Trestles / Working at Height

People falling from heights is the primary cause of death on construction sites in Ireland, where 12 people fell to their deaths in 2004.

Hazards → Risks

Collapse of the scaffolding, due to poor design or construction or overloading	**Serious injuries, especially back injury / paralysis**
Incomplete scaffold	**Fractures**
Unauthorised alterations	**Death**
Falls from height	

→

Controls →
Managing the Risks

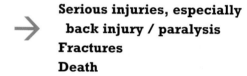

* Select the most suitable scaffold for the task.
* Allow only a competent scaffold company / competent persons to erect, alter and dismantle scaffolding, as per the Code of Practice.
* Ensure that every scaffold, and every part thereof, is of good design and construction, of suitable and sound material and of adequate strength for the purpose for which it is to be used.
* Do not use any defective material / part for a scaffold.
* Ensure boards are free from defects and arranged to avoid tripping or tipping.
* Ensure that all sole-plates, base-jacks, etc are in good condition and suitable for the task. Base-jack is only required where a concrete surface is used - the base-jack must sit on the sole-plate, ensuring that no part of the base-jack over-hangs the sole-plate.
* Ensure that working platforms are fully boarded.
* Ensure that loading platforms are clearly labelled with the SWL - for example, 'Max. SWL (2.5 bay) 790kg = 30 blocks + 2 workers'.

* Erect scaffold on a firm / even surface and use adequate base-plates / sole-plates.
* Protect all exposed steel bars (reinforcing) suitably.
* Provide hand-rails of adequate strength at every lift, not less than 950mm above the platform. Use double hand-rails on the outer face.
* Do not leave scaffolding partly-erected. Use warning signs, where necessary.
* Do not allow workers / sub-contractors to interfere with scaffolding, or to remove toe-boards / guard-rails / boards, etc.
* Do not allow workers / sub-contractors to use a scaffold, unless it is complete and fit for work. If in doubt, inspect the most recent CR 8 / Scafftag system, and check with Site Manager.
* Ensure all CR 8s are completed weekly by a competent person and kept in the site office.
* Report any defect in scaffolding immediately to supervisor / site management / safety officer.
* Remove all materials, tools and equipment from scaffold as soon as work is completed. This is the responsibility of each sub-contractor / worker.
* Do not fix tarpaulin sheets (or other windsails) to a scaffold unless it has been specifically designed to take them.
* Where a doubt exists as to the safety of a scaffold, DO NOT USE. Seek guidance from site management.

Mobile Scaffold Towers
Ensure mobile scaffold towers / platforms:
* Are constructed and braced properly.
* Have a fully-boarded platform.
* Have toe-boards and hand-rails.
* Have a secured ladder access.
* Are used on suitable, level ground.
* Have adequate height-to-base ratio.
* Have the stabiliser secured.
* Have wheel nuts locked.

Use a recommended height to base ratio of 3 : 1 maximum.

Never move a mobile scaffold with workers or materials on top.

Never use an incomplete scaffold.

Safe Working Loads

Ensure that SWLs are assigned as per manufacturer's guidelines, by a competent and experienced person. Do not exceed SWLs. Use sample weights for calculating SWLs as shown below.

Sample weights	KG
Person (average)	80
Wheelbarrow full of mortar	150
Board and Mortar	30
500 concrete bricks (15N / mm)	1,750
50 concrete blocks	1,020

Applicable Legislation / Standards / Code of Practice

* SHWW (General Application) Regulations, 1993, Part IV: Work Equipment Regulations.
* SHWW (Construction) Regulations. 2001, Part 13: Working at Heights, Regs.51-62 & 66-71.
* SHWW Act, 2005.
* *Proposed SHWW (Construction) Regulations, 2006.*
* *Proposed SHWW (General Application) Regulations, 2006, Part XV: Working at Height.*
* HSA: *Code of Practice for Access & Working Scaffold.*

> **Under the Construction Regulations, 2001, scaffolding must be inspected:**
> - **Before being taken into use.**
> - **After any modification, period without use, exposure to bad weather or any other circumstances that may have affected its strength or stability or displaced any part of it.**
> - **At least once every seven days - record of inspection to be kept on CR 8 forms.**

Training / Certification

* CIF - Scaffolding Inspection Course - designed for construction personnel who manage, and who are required to perform, simple scaffolding inspections and form an opinion regarding its safety.
* FÁS CSCS - Basic / Advanced Scaffolding Courses.

Further Information

* HSE: *General Access Scaffolds & Ladders*, CIS No.49(rev1).
* HSE: *Tower Scaffolds*, CIS10(rev3).
* NASAC - National Association of Scaffolding & Access Contractors.

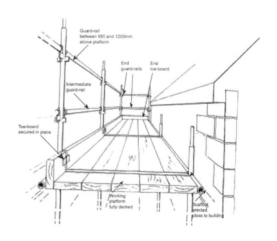

Cherry picker (MEWPs) - see p.152

SHARPS

What are sharps?

Sharps include hazards such as protruding nails.

Hazards → Risks

Broken glass / nails / edges **Lacerations**
Blood loss / infections

Controls → Managing the Risks

* Ensure that workers wear appropriate puncture-protective footwear and use gloves when gathering glass / sharps / nails.
* Do not use timber / material with projecting nails in any work, if such timber or material is a source of danger to persons.
* Make safe all scrap timbers with protruding nails.
* Wear protective footwear at all times on site.

Applicable Legislation

* SHWW (General Application) Regulations, 1993.
* SHWW (Construction) Regulations, 2001, Part 4: Reg.18.
* SHWW Act, 2005.

Sharps can cause serious injury - put them in separate timber skips immediately

SILICA

See also: Dust

What is silica?

Silica occurs as a natural component of many materials used or encountered in construction activities. Crystalline silica is present in substantial quantities in sand, sandstone and granite, and often forms a significant proportion of clay, shale and slate. Products such as concrete and mortar also contain crystalline silica.

Hazards → Risks

Stone masonry, façade renovation, blast-cleaning with sand, demolition, concrete scabbling, cutting or drilling, tunnelling

Breathing difficulties, scarring of the lung tissue, progressive silicosis (this causes fibrosis - hardening or scarring - of the lung tissue, with a consequent loss of lung function)

Controls → Managing the Risks

* Eliminate use of silica - for example, use non-silica grits for blasting.
* Always use exhaust-ventilated tools that remove the dust at source and tools fitted with a water supply for dust suppression.
* Wear appropriate PPE & RPE at all times.
* Ensure that all RPE is selected by a competent person and provide training in its use to all workers.
* Provide respiratory questionnaires, lung function testing and chest x-rays to high-risk workers.

Applicable Legislation

* SHWW (General Application) Regulations, 1993.
* SHWW (Chemical Agents) Regulations, 2001.
* SHWW (Construction) Regulations, 2001.
* SHWW Act, 2005.
* *Proposed SHWW (Construction) Regulations, 2006.*
* *Proposed SHWW (General Application) Regulations, 2006.*

Further Information

* HSE: *Silica*, CIS36(rev.1).

Silica has been assigned a maximum exposure limit (MEL) of 0.3 mg / m3, expressed as an 8-hour time weighted average (TWA).

SITE SECURITY

See also: Fencing / Lone Working / Sub-contractors / Public Safety

Hazards → Risks

		Injuries
Unauthorised use		**Lacerations**
of equipment	→	**Fractures**
Entrapment		**Sprains / strains**
		Crush injuries
		Death

Controls → Managing the Risks

* Before any works start at any location, carry out a Risk Assessment to identify the hazards posed by trespass / vandalism on that particular site and the risks arising.
* Put in place the appropriate control measures (fencing / barriers / locks, etc.), as far as is reasonably practicable.
* Immobilise /secure all machinery / plant / equipment, when not in use and at the end of the working day, so as to prevent unauthorised use / access.
* Secure the site, as far as is reasonably practicable, to prevent unauthorised access.
* Erect a warning sign at each entrance, warning persons of the dangers and denying unauthorised access.
* Report all unauthorised access immediately to site management.
* Ensure the safe removal off the site of any person who gains unauthorised access.
* Review the effectiveness of site security arrangements in the light of experience. In particular, review their adequacy if there is evidence of children playing on, or near the site.

How access is controlled depends on the nature of the project, the risks and the location. Physically define the boundaries of the site, where practical, by suitable barriers that reflect the nature of the site and its surroundings. In deciding on the most appropriate exclusion methods to prevent unauthorised persons entering the site, consider:

* The location of the site: Is it located in an urban area and close to extensive housing or a school, or is it in a remote area?
* Is there a right of way across the site?
* Have the public or others access to the site in, or next to, other work areas?

* If it is not practicable to erect hoarding around the site, can hazardous areas be cordoned off?
* Are new houses being built on a development where some houses are already occupied?
* Are there are children and other vulnerable people nearby?
* What is the nature of the work and the risk to persons not authorised to enter the site?

External Security Companies on-site
Security companies working on construction sites are subject to the same conditions as any other sub-contractor:

* They must receive Induction training and comply with site rules.
* They must submit the following documentation:
  o Safety Statement.
  o Training Records - Safe Pass and Security Institute qualifications.

Applicable Legislation

* SHWW (General Application) Regulations, 1993.
* Occupiers' Liability Act, 1995.
* SHWW (Construction) Regulations, 2001.
* Private Security Services Act, 2004.
* SHWW Act, 2005.
* *Proposed SHWW (Construction) Regulations, 2006.*
* *Proposed SHWW (General Application) Regulations, 2006.*

Further Information

* Security Institute of Ireland - www.sii.ie - Tel: 01-454 0439.

SLIPS & TRIPS

See also: House-keeping

Hazards / Risks

See below.

Controls → Managing the Risks

Typical Slip Hazards	Managing the Risk
Spills of wet / dry substances	Design the process to reduce or contain spillages - for example, by providing lips around tables and bunds around equipment. Clean the spillage immediately with a suitable cleaning agent. Use signs to let people know the floor is wet and provide alternative routes around the wet area.
Moving from a dry to a wet surface - for example, inside doorways	Place doormats inside external doorways to reduce tracking of water into building. Warn of slip hazard with signs. Ensure suitable footwear is used.
Slippery surfaces	Use cleaning methods that do not create slippery surfaces. Use stick-on strips or mats. Abrade or treat the surface chemically. Remove build-ups of ice regularly. Grit or salt iced areas as appropriate.
Footwear	Ensure that people entering the area wear slip-resistant footwear suitable for the type of work.

Typical Trip Hazards	Managing the Risk
Rubbish	Provide convenient bins for placing rubbish into. Clear rubbish promptly. Keep walkways clear.
Materials stored or left in passageways	Provide adequate storage space and shelving for stock. Mark passageways and keep the full width of passageways clear.
Trailing leads / cables	Place equipment and power outlets to avoid need for trailing cables.
Damaged / torn flooring	Promptly remove and replace damaged / torn flooring.
Rugs / mats	Securely fix rugs / mats to prevent them slipping on the surface beneath, and so that edges are prevented from curling.
Changes of slope / level	Provide good lighting and remove / highlight changes in level. Provide secure non-slip floor markings, particularly on stair treads, steps and changes in slope. Provide handrails following slope.
Lips caused by weather-stripping etc. at doorways	Design doorways to avoid need for lips. Provide slope up to lip. Provide markings and improved lighting.
Hidden hazards due to poor lighting	Provide adequate lighting for all work areas so that dark or shadowed areas are avoided.
Uneven surfaces, bumps and holes	Provide working areas with even surfaces, free from bumps and holes. Provide alternative routes to avoid uneven ground. Repair holes in surfaces promptly.

Applicable Legislation

* SHWW (General Application) Regulations, 1993, Part II.I
* SHWW (Construction) Regulations, 2001.
* SHWW Act, 2005.
* *Proposed SHWW (Construction) Regulations, 2006.*
* *Proposed SHWW (General Application) Regulations, 2006.*

SOLVENTS

See also: Chemicals

What are solvents?

Solvents are chemical substances. In construction products, they act as carriers for surface coatings such as paints, varnishes, adhesives and pesticides. The most common solvents found in construction are:

* White spirit - in paints, varnishes, cleaning products.
* Xylene - in paints, adhesives, pesticides.
* i-butanol - in natural and synthetic resins, paints, lacquers.

Hazards → Risks

Exposure to solvents → **Irritation of the skin, eyes, lungs**
Headache, nausea
Dizziness, light-headedness
Impaired co-ordination
Unconsciousness

Controls → Managing the Risks

* Ensure work area is well-ventilated, where work is being carried out with solvents.
* Store solvents in properly-labelled, suitable containers. Keep lids on containers unless contents are being poured or dipped, etc.
* Use sealed containers for solvent waste.
* Ensure all workers are trained in the dangers of solvents, how to minimise exposure and in how to deal with spillages.
* Do not smoke in areas where solvents are being used.
* Wear appropriate PPE when working with solvents.
* Maintain a high standard of personal hygiene to avoid contamination.

Emergency First Aid measures
* Wash solvent splashes off the skin with running water (10 minutes).
* When washing solvents from the eye, always turn affected eye down, so not to infect the good eye.
* Remove heavily-contaminated clothing as soon as possible.
* Cover wounds with suitable dressing.
* Remove to hospital ASAP.

Hygiene measures

* Provide washing / changing facilities on site (main contractor's responsibility).
* Encourage all workers to wash their hands before eating, drinking, smoking and after going to the toilet.
* Provide canteen facilities for the consumption of food.
* Make available hand cleaner / barrier creams for workers.

Applicable Legislation / Code of Practice

* SHWW (General Application) Regulations, 1993.
* SHWW (Chemical Agents) Regulations, 2001.
* SHWW Act, 2005.
* *Proposed SHWW (Construction) Regulations, 2006.*
* *Proposed SHWW (General Application) Regulations, 2006.*

* HSA: Code of Practice for the SHWW (Chemical Agents) Regulations, 2001.

Further Information

* HSA: *Chemical Legislation: An Overview*.
* HSE: *Chemical Cleaners*, CIS24(rev1).
* HSE: *Solvents*, CIS27(rev2).

NOTES

STEEL FIXING

See also: Abrasive Wheels / Pre-cast Elements & Components

Hazards → Risks

Falls from heights
Slips and trips
Abrasions from bar ends /
  tie wires

Head injuries
Lacerations
Abrasions
Impalement

Controls → Managing the Risks

* Lay out reinforcement before fixing, so that individual bars may be easily located.
* Ensure that all workers cutting reinforcement have received Abrasive Wheel training and have suitable experience.
* Wear suitable PPE at all times, when handling and cutting reinforcement.
* When working on a previously-cast suspended slab, ensure that supporting formwork is designed to support additional weight.
* Put adequate guards on bar-bending machines and ensure that all guards are in place at all times.
* Ensure all bars are bent in accordance with BS 4466.
* Ensure that spacers are strong enough and at suitable centres to carry the weight of the reinforcement.
* Provide walkways, where necessary - access over steel cages, etc.
* Fit 'mushroom' cap / box over all projecting steel reinforcement.
* Never lift reinforcement bundles by their binding wire.
* Lift only one sheet at a time when handling fabric.
* When lifting fabric by a crane, always use four biting points.

Applicable Legislation / Standards

* SHWW (General Application) Regulations, 1993, Part IV: Work Equipment Regulations.
* SHWW (Construction) Regulations, 2001.
* SHWW Act, 2005.
* *Proposed SHWW (Construction) Regulations, 2006.*
* *Proposed SHWW (General Application) Regulations, 2006.*

* BS 4466: *Specification for Bending Dimensions & Scheduling of Specification for Concrete.*

Further Information

* Irish Steel Fixing Association.

STRESS

See also: Bullying / Harassment

What is stress?

Stress is the adverse reaction people have to excessive pressure. It is not a disease but, if stress is intense and long-term, it can result in mental and physical ill-health.

Signs of stress include: changes in a person's mood or normal behaviour, deteriorating relationships with co-workers, irritability, indecisiveness, absenteeism and reduced performance.

Problems that can lead to stress include:

* Lack of communication and consultation.
* A culture of blame when things go wrong, denial of potential problems.
* An expectation that people will regularly work excessively long hours.
* Having too much to do / too little time.
* Being under- / over-qualified for the job.
* Boring / repetitive work.
* The physical work environment - noise, harmful substances, threat of violence.
* Lack of control over work activities.
* Bullying, racial or sexual harassment.
* Fears about job security.
* Confusion about how everyone fits in.
* Lack of support from management.

Hazards → Risks

Problems as above **Dependency on medication**
Sleeplessness
Ill-health
Excessive smoking and drinking

Controls → Managing the Risks

* Prepare a policy statement to outline the procedures to be taken if stress is identified as a hazard. Make available the procedures to management and employees.

* Where an employee is identified as suffering from stress, carry out a Risk Assessment, identify the stressors (see problems that lead to stress) and take the appropriate corrective measures.

* Through good staff communication, make clear that management supports all such staff members and is prepared to make organisational changes, where necessary.

* If the source of the stress is of a personal nature, encourage employees to contact the Safety Officer or a member of management.

Applicable Legislation

* SHWW (General Application) Regulations, 1993.
* SHWW Act, 2005.
* *Proposed SHWW (Construction) Regulations, 2006.*
* *Proposed SHWW (General Application) Regulations, 2006.*

Further Information

* HSA: *Work Positive* pack (includes CD).
* HSA: *Work-related Stress: A Guide for Employers*.
* HSE: *A Short Guide to Work-related Stress*, INDG281(rev1).
* HSE: *Tackling Work-related Stress: A Managers' Guide to Improving & Maintaining Employee Health & Well-being*, HSG218.
* Oak Tree Press: *Surviving Stress: A Guide for Managers & Employees*, Samuel Malone (2004).

SUB-CONTRACTORS

See also: Site Security

Hazards → Risks

Lack of communication   **Poor working**
Conflicting work schedules   **environment**
Operations that endanger   →   **Accidents**
  other workers   **Injuries**

Controls → Managing the Risks

Contractors and self-employed persons must:

* Provide their Safety Statement and Method Statements to PSCS, before starting work on site.

* Ensure that all relevant persons under their control have received Safe Pass and CSCS training and have been issued with registration cards. Written documentation is required to support this.

* Provide evidence showing that appropriate Employers / Public Liability insurance is in effect for the site.

* Bring to the attention of the site foreman / Safety Officer, and anyone else who may be affected, any work process / use of materials that may endanger health and safety while at work.

* Comply with the requirements of the site Safety & Health Plan and co-operate with site management in providing a safe place of work, safe system of operation, competent employees and the wearing of protective clothing / equipment.

* Ensure that their employees, and others under their care, are in possession of appropriate PPE and will wear such PPE as required.

* Report any defect in the plant / equipment / place of work without delay.

* Attend a pre-start meeting with the PSCS / Safety Officer where other site-specific rules will be issued.

* Attend Induction training before starting work and attend weekly / fortnightly Toolbox Talks held on-site.

* Assist in all stand-downs / investigations due to a breach of safety on site.

* Not sub-let work without the written permission of PSCS / main contractor. It is the responsibility of the sub-contractors and self-employed persons to ensure that the required level of insurance cover is maintained by any person(s) or sub-contractor(s) to whom they sub-let work. Evidence of such insurance cover and a Safety Statement must be produced to the PSCS / main contractor before their consent will be given to the sub-letting.

* Comply with the provisions of the Construction Regulations, 2001, Reg.10, and inform the PSCS / Safety Officer in writing of the name of the nominated Safety Officer / Safety Representatives.

* Report all accidents to the site foreman / Safety Officer, no matter how trivial, in writing. If any sub-contractor's personnel has an accident on site that requires medical attention or involving lost time, then a full report of the circumstances, including witnesses' statements, must be submitted within 24 hours.
* Wear ear protection when work is being carried out in an area where the noise level exceeds 85 DBA. When the source of the noise is from sub-contractor's equipment, the sub-contractor must identify the area with signs and supply suitable ear protection to his employees.

> **Co-operation**
> **+**
> **co-ordination**
> **=**
> **SAFE SITE**

Any person found not complying with the above or any other safety directive should not be permitted to continue working on the site.

Before selecting a contractor, ask:
* What experience have they in the type of work?
* What are their health and safety policies and practices?
* About their recent health and safety performance (number of accidents, etc).
* What qualifications and skills have they?
* For their Safety Statement and Method Statement.
* What health and safety training and supervision they provide?
* Their arrangements for consulting their workforce.
* Whether they have any independent assessment of their competence?
* Whether they are members of a relevant trade or professional body (for example, CIF / NISO)?

Applicable Legislation

* SHWW (General Application) Regulations, 1993.
* SHWW (Construction) Regulations, 2001: Part III, Regs. 9-14; Fourth, Fifth and Eighth Schedule.
* SHWW Act, 2005.
* *Proposed SHWW (Construction) Regulations, 2006.*
* *Proposed SHWW (General Application) Regulations, 2006.*

Further Information

* HSA: *Workplace Health & Safety Management.*
* HSE: *Managing Contractors: A Guide for Employers*, HSG159.
* HSE: *Use of Contractors*, INDG368.
* HSE: *Working Together: Guidance on Health & Safety for Contractors & Suppliers*, INDG268(rev).

SUN DAMAGE

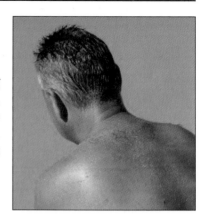

Sun damage causes the most common form of cancer in Ireland, with 8,000 cases every year. 90% of skin cancers are caused by exposure to sun rays and are preventable by taking proper precautions. Construction workers are a high-risk category but particularly at risk are people who have:

* Pale, freckled skin that does not tan, or burns before it tans.
* Red or fair hair & light coloured eyes.
* Large number of moles (50 or more).
* History of sunburn / skin cancer.

Hazards → Risks

Over-exposure to sun
Outdoor work

→

**Immediate effects - skin reddens, blisters, skin colour deepens
Delayed effects - sunburn (four hours later), upper layers of skin peel, sunstroke
Long-term effects - wrinkles, premature aging, cancer may develop**

Controls → Managing the Risks

* Limit sun exposure as much as possible.
* Aim to avoid direct sunlight between 10am-3pm whenever possible.
* Train workers periodically in the prevention / recognition / treatment of sunburn / heatstroke / heat exhaustion.
* Do not allow workers to work bare-chested or in shorts.
* Ensure that workers drink lots of cool liquids when working in heat.
* Use appropriate barrier creams used on exposed areas of skin.

First Aid measures:
* Rest the casualty in a cool, shaded area.
* Encourage the casualty to drink lots of cool water.
* Loosen all tight clothing.
* Use wet towels / clothing to aid cooling.
* If you suspect heat stroke, call an ambulance immediately and alert a First Aider.

Further Information

* Irish Cancer Society.
* Construction Workers' Health Trust.

Cranes & Lifting operations - see p. 84 / 136

TELEPORTERS

See also: Chains & Slings / Forklifts / Lifting Equipment & Operations / Machinery - General

Hazards → Risks

Lifting
Over-turning
Falling objects   →
Over-loading
Reversing

Serious injuries
Crushing
Unconsciousness
Death

Controls → Managing the Risks

* Operate and maintain teleporters as per manufacturer's instructions.
* Do not exceed load restrictions.
* Allow only competent / certified workers to carry out work on teleporters.
* Remove the keys when the teleporter is not in use and park it so as not to present a danger to anyone.
* Do not operate a teleporter while under the influence of alcohol / drugs, including prescribed drugs.
* No unauthorised riding on the teleporter unless it is designed for same.
* Ensure that operators always wear a safety belt while inside the machine and that they keep the lower portion of the door closed.
* Ensure that drivers remain inside the cab in the event of the machine over-turning.
* Operate controls from inside the safety of the cab only.
* Do not use mobile phones when operating a teleporter.

* Do not wrap chains and slings around the forks of a teleporter when it is used to lift loads.
* When using chains or slings with forks, use suitable fork clamps, with the chain or sling suspended from a suitable hook or shackle.
* When lifting with a teleporter, remove the forks and use a crane extension with hook or shackle.
* Ensure that all teleporters are checked by a competent person every 14 months.

Applicable Legislation

* SHWW (General Application) Regulations, 1993, Part IV: Work Equipment Regulations.
* SHWW (Construction) Regulations, 2001, Part 11, Transport, Earthmoving & Materials Handling Machinery, Regs.41-46.
* SHWW Act, 2005.
* *Proposed SHWW (Construction) Regulations, 2006.*
* *Proposed SHWW (General Application) Regulations, 2006.*

Training / Certification

* FÁS - CSCS Telescopic Handler operation.

Required Documentation

* CR4A - Lifting Appliances - Report of thorough examination (14 months / repair / first use).
* CR4B - Lifting Appliances - Weekly Inspection report.

TIMBER FRAME

See also: Scaffold

Although still relatively new to the Irish market, timber frame has quickly taken over from traditional block housing.

Hazards → Risks
See below.

Controls → Managing the Risks

Activity	Hazard	Controls → Managing the Risks
Site Preparation	Inadequate access / unstable ground	Prepare a suitable site entrance to accommodate all site transport / delivery vehicles. Clear / compact all work areas / access routes to permit safe movement of vehicles / personnel.
Preparing / erecting timber frame houses	Lack of knowledge	Train all workers in the safe erection of the houses and ensure that they are familiar with all documentation issued by the kit manufacturer.
Construction of Block Party Wall	Free-standing structure	Brace the wall with suitable supports as it is constructed. Erect the bracing as per the *Timber Frame Guidelines*. Ensure the bracing is done to a structural engineering design.
Erection of houses	Scaffold	Ensure all scaffold is erected by competent persons as per the *Code of Practice for Access Scaffold* and as per the *Timber Frame Guidelines*. Ensure all scaffold is inspected before use by a competent person. Ensure all documentation (CR 8 / Handover Certificates) is completed by competent persons and kept on site. Erect scaffold before the erection of the timber frame kit, on all sides. Use handrails on the inside of ladders and loading bays.

Activity	Hazard	Controls → Managing the Risks
Crane / Lifting Equipment	Falling objects	Use only cranes capable of handling the required capacity. Ensure that all equipment / operators satisfy current legislative requirements as per Appendix D of the *Guidelines*. Mark exclusion zones clearly and ensure a competent banksman is available to assist the crane operator.
Crane / Lifting Equipment	Unsuitable weather	Confirm that the crane / lifting equipment operator is the sole decision-maker on unsuitable / unsafe weather for operating safely.
Crane / Lifting Equipment	Unloading	Ensure that loads delivered to site do not exceed 4.25 metres. Ensure that all unloading is done with the use of correctly secured and erected ladders. Unload all wall panels using the crane / lifting equipment.
Erection of Ground Floor Panels	Swaying / unsecured panels	Use a suitable rope to guide the panels into position. Hold the wall panel in place with the crane / lifting equipment until it is fixed / secured in position. Ensure that all workers using shot-firing equipment are trained / competent and that they wear the appropriate safety goggles and hearing protection, etc.
Positioning of Floor Panels	Unsecured panels	Position all floor panels using the crane / lifting equipment. Use inertia reel harnesses (as per the manufacturer's instructions and site-specific Risk Assessment), if a worker is required to go out onto already-positioned floor panels.
Erection of First Floor Panels	Unsecured panels	Use a suitable rope to guide the panels into position. Hold the wall panel in place with the crane / lifting equipment until it is fixed / secured in position. Fix all panels permanently before moving on to the next stage of the erection.

Activity	Hazard	Controls → Managing the Risks
Erection of the Roof	Working at height	Use an appropriate fall protection system (netting) where the fall distance is greater than 2 metres. Ensure that scaffolding / fall protection systems, once installed, are inspected by a competent person before work commences. When spreading trusses, ensure all workers use the scaffold.
Felt and battening of the roof	Working at height / openings	Leave all safety systems in position until felt and battening is completed.
Fixing of the Kit	Unsecured panels / straps	Fix the entire structure as per the manufacturer's on-site nailing instructions, including anchor straps. Check all fixings to ensure that the schedules are completed.
Brace removal	Sharps / nails	Take care when removing temporary bracing to ensure that all nails are removed.

Timber Frame Checklist	**Signed / Dated**
Training / Toolbox Talk	
Site Preparation	
Construction of block party wall	
Scaffolding	
Crane / Lifting equipment	
Unloading of trailer	
Erection of the ground floor wall panels	
Positioning of the floor panels	
Erection of the first floor wall panels	
Installation of protection systems	
Erection of the roof	
Felt and battening of roof	
Fixing of kit	

Applicable Legislation

* SHWW (General Application) Regulations, 1993, Part IV: Work Equipment Regulations.
* SHWW (Construction) Regulations, 2001.
* SHWW Act, 2005.
* *Proposed SHWW (Construction) Regulations, 2006.*
* *Proposed SHWW (General Application) Regulations, 2006.*

Further Information

* HSA: *Timber Frame Erection Guidelines. Method Statement for Kit Erection.*

* Irish Home Builders Association.
* Irish Timber Frame Manufacturers' Association.

NOTES

TRESTLES

See also: Scaffolding / Working at Height

Hazards → Risks

Falls   →   **Injury / crushing / death**

Controls → Managing the Risks

✶ Ensure that trestles are erected only by competent persons according to the manufacturer's guidelines.

✶ Inspect trestles for defects / faults, etc, before use.

✶ Do not allow workers to use trestles unless they are safe for working on.

✶ Do not use trestles on unstable surfaces. Check ground conditions before erection.

✶ Trestles are work platforms designed for low-level work - do not use for work over 2m.

✶ Only use trestles in a single tier - do not place one trestle on another to gain height.

✶ Use only 'locating pins' recommended by the manufacturer and take care at all times when inserting locating pins.

✶ Do not use trestles on scaffold unless the width is a minimum of 1,050mm and a minimum of five boards.

✶ Ensure that the platform level on trestles is fully boarded. Do not allow workers to remove boards without prior consent from site management.

✶ Do not over-load platforms.

✶ Use suitable fall protection at all times.

Applicable Legislation

✶ SHWW (General Application) Regulations, 1993, Part IV: Work Equipment Regulations.

✶ SHWW (Construction) Regulations, 2001, Part 13: Working at Heights, Reg.64 (1-3).

✶ SHWW Act, 2005.

✶ *Proposed SHWW (Construction) Regulations, 2006.*

✶ *Proposed SHWW (General Application) Regulations, 2006, Part XV: Working at Height.*

Further Information

✶ Manufacturers' guidelines.

✶ HSE: *Working at Heights Regulations: A Brief Guide*, INDG401.

Scissors Lift (MEWPs) - see p. 152

TUNNELLING

See also   Confined Spaces / Excavations

Hazards → Risks

Falls
Inundation   **Injury / crushing**
Confined spaces   →   **Drowning**
Gas inhalation   **Death**
Collapse of tunnel

Controls → Managing the Risks

* Before any excavation operation begins, inspect and survey the site to ascertain the condition of structures, ground and services, including overhead / underground lines / gas pipes.

* Before excavating, inspect the adjacent area to ensure that digging operations will not cause other structures to become unstable or collapse.

* Where required, use underpinning and propping to stabilise all adjacent structures that may be affected by excavation / tunnelling operations.

* Ensure that all excavations (more than 2m deep), earthworks, etc are inspected / controlled by a competent person as per Regulations.

* Ensure that Risk Assessments, Method Statements and Permits to Work are all in place for work in confined spaces / tunnels, and that they are reviewed and approved by the PSCS / Safety Officer.

* Allow only responsible / competent persons to work in tunnels, and train all workers in the immediate hazards.

* Provide safe access / egress into the excavation / tunnel.

* Ensure that designed pedestrian walkways are in place to separate site traffic from people entering / leaving tunnels.

* Ensure that adequate / sufficient propping is in place to prevent the collapse of the walls and ceiling of the tunnel.

* Monitor oxygen level and harmful gases at all times. A supply of fresh air may be required, depending on the depth of the tunnel. Seek advice from a competent person.

* Install sufficient lighting, including emergency lighting, in any tunnels for work operations, access, egress, etc.

* Ensure that a 'buddy' system is in place when working in tunnels. Do not allow any worker to work alone.

* Ensure that a suitable communication system is in place at all times, to allow verbal communication between those on the surface and those in the confined area.

* Wear suitable / approved PPE at all times.

* Test all RPE before use.
* Develop emergency / rescue procedures and train all workers in these procedures before working in tunnels.

> Tunnelling is extremely dangerous.
> It is specialised work that should not be undertaken without relevant experience / supervision.

Applicable Legislation

* SHWW (General Application) Regulations, 1993.
* SHWW (Construction) Regulations, 2001.
* SHWW Act, 2005.
* *Proposed SHWW (Construction) Regulations, 2006.*
* *Proposed SHWW (General Application) Regulations, 2006.*

Training

* FÁS - Confined Space Entry Training.

Further Information

* HSA: *A Guide to Safety in Excavations*.
* HSA: *Working in Confined Spaces*.

NOTES

VIBRATION

See also: Hand & Power Tools / Noise

What causes vibration?

* Chainsaws.
* Jack-hammers / rock-breakers / concrete-breakers / road-drills.
* Hammer drills.
* Hand-held grinders and sanders.
* Power hammers and chisels.
* Riveting hammers and bolsters.

Vibration: cause and effect

Hazards → Risks

Vibration White Finger, characterised by attacks of whitening when the fingertips become numb, is caused by damage to blood circulation. Other damage may be to the nerves and muscles of the fingers and hands, causing numbness and tingling, reduced grip strength and sensitivity.

Signs & symptoms

* In cold or wet weather, fingers go white, then blue and then red when warming up and are very painful.
* You cannot feel things with your fingers.
* You have difficulty picking up small objects such as nails / screws.
* Pain, tingling, numbness in hands, wrists and arms.
* Loss of strength in hands.

Users of rotating or percussive hand-guided tools, where the hands are exposed to high levels of vibration, are at greatest risk. The degree of risk depends on:

* The amount of tool vibration.
* The length of time the tool is used for.
* Intermittent or continuous use.
* Workplace temperature.
* Individual susceptibility.
* Method of work.

Controls → Managing the Risks

* Allow only authorised / competent persons to operate plant / machinery.
* Instruct all workers in the potential sources and health effects of vibration.

* Wear appropriate hearing / eye / hand protection. Provide training in the correct use and care of PPE.
* Maintain all machinery as per manufacturer's instructions, particularly suspension seats / components.
* Identify the vehicles / machines / work situations with the highest level of vibration and arrange a rota for operators / drivers to reduce the time spent on them.
* Where possible, construct jogs to hold materials or tools.
* For pneumatic machinery, wear protective gloves to help prevent vibration white finger.

A gravel-flattener

Selecting low-vibration tools

* Tools with CE mark are declared by the manufacturer to be safe when used as instructed.
* Manufacturers identify vibration levels in m/s2.
* Vibration data is given in technical sales literature and the instruction book, if the vibration level exceeds 2.5 m/s-2 during standard tests.
* Supplementary information on measures necessary to control risks from exposure to tool vibration should appear in the instruction book.
* Low-vibration tool accessories should be selected.

Applicable Legislation / Standards

* SHWW (General Application) Regulations, 1993, Part IV: Work Equipment Regulations.
* SHWW (Construction) Regulations, 2001.
* SHWW Act, 2005.
* *Proposed SHWW (Construction) Regulations, 2006.*
* *Proposed SHWW (General Application) Regulations, 2006.*

> **Note: The proposed SHWW (General Application) Regulations, 2006, Part XVI: Control of Vibration of Work, is entirely new to Irish law and will introduce the provisions of the EU Vibration Directive (2002/44/EC). There may be delays in bringing rules regarding vibration into force, as the proposed Regulations take full advantage of the transitional periods allowed the EU Directive (up to July 2007).**

* BS / EN / ISO 8662-14:1997. *Hand-held Portable Power Tools - Measurement of Vibrations at the Handle. Part 14: Stone Working Tools & Needle-scalers.*

Further Information

* HSE: *Reducing the Risk of Hand-Arm Vibration Injury among Stonemasons*, MISC112.
* HSE: *Hand-Arm Vibration*, HSG88.

VISUAL DISPLAY UNITS

See also: Office Work

Regulations govern the use of VDUs:

* Where employees have no discretion as to the use or non-use of VDUs.
* Where VDUs are normally used by employees for periods of more than one hour at time.
* Where VDUs are used by employees daily.

On-site engineers, architects, project managers and clerical staff may be classified as users.

Hazards → Risks

Poor posture
Inadequate seating
Poor design
Inadequate lighting /
   screen glare

Physical musculoskeletal problems / upper limb pains
Eye fatigue
Mental stress

Controls → Managing the Risks

* Design the work activities of VDU users so as to give periodic breaks away from the screen.
* Carry out annual VDU workstation assessments and keep records of same (employer's responsibility).
* Provide suitable eye / eyesight tests for users, before the user starts work and at regular intervals (employer's responsibility).
* Train all VDU users in the safe use of the workstation, including induction training, principles of ergonomics, furniture adjustment, screens, keyboards, lighting, etc.
* Train workers in best working postures / procedures and recognition of potential strains.
* Limit the use of laptops. Ensure that all intricate work, such as developing CAD drawings, is done at computer workstations.

Screen:
* Must display clearly-defined characters of adequate size.
* Must have adjustable brightness and contrast.
* Must have adjustable swivel and tilt.
* Must be positioned free from reflective glare.

Keyboard:
* Should be made from non-reflective material.
* Must have clearly-defined characters.
* Must have sufficient space in front of the keyboard to provide support to employee's hands and arms.

Worktop:
* Must be sufficient size for work and provide enough room to allow employee to work.
* Worktops should be made of a non-reflective surface.

Chair:
* Must be stable.
* The seat and back should be adjustable in height, and the back in tilt.
* Provide a footrest for comfort to those who request it.

Environment:
* The immediate environment should have sufficient space and lighting.
* Temperature and humidity values should be maintained at optimal levels.
* Noise levels should be monitored.

Applicable Legislation

* SHWW (General Application) Regulations, 1993, Part IV: Work Equipment Regulations; Part VII: Display Screen Equipment (VDUs) Regulations; Second, Third & Fourth Schedules.
* SHWW (Construction) Regulations, 2001, Fourth Schedule.
* SHWW Act, 2005.
* *Proposed SHWW (Construction) Regulations, 2006.*
* *Proposed SHWW (General Application) Regulations, 2006, Part VI: Display Screen Equipment.*

Further Information

* HSA: *VDU Regulations - An Easy Guide for Employees.*
* HSA: *Guidelines to The SHWW Act, 1989 & The SHWW (General Application) Regulations, 1993* (p.116-131) - good pictorial examples.
* HSE: *Officewise.*
* HSE: *Working with VDUs.*

WELDING

See also: Acetylene / Chemicals

Hazards → Risks

Careless handling of a lighted
  blowpipe resulting in burns
  to the user or others

Using the blowpipe too close
  to combustible material

Cutting up or repairing tanks or
  drums that contain or may have
  contained flammable materials

Gas leaking from hoses, valves
  and other equipment

Misuse of oxygen

Backfires / flashbacks

Burns

**Eye damage from
  metal fragments,
  sparks, etc**

**Fire damage,
  due to accidental ignition**

Controls → Managing the Risks

✶ Use a 'Permit to Work' system for all welding.

✶ Ensure all oxyacetylene equipment has a flashback flame arrestor and a
  non-return valve.

✶ Ensure visual pressure gauges / volume indicators are fitted.

✶ Inspect welding equipment regularly (particularly the welding tip and
  hosing) for signs of wear.

✶ Never weld wheels with tyres fitted.

✶ Always turn the gas supply off at the cylinder when the job is finished.

✶ Keep hoses clear of sharp edges / abrasive surfaces or where vehicles can
  run over them.

✶ Check all connections and equipment regularly for faults and leaks.

✶ Always provide adequate ventilation during welding.

✶ Store gas cylinders outside, whenever possible, or in a well-ventilated
  place.

✶ Avoid taking gas cylinders into poorly-ventilated rooms or confined
  spaces.

✶ Secure all acetylene cylinders in the upright position and protect from
  damage in racks or trolleys.

✶ Train workers in safe working procedures and provide suitable protective
  equipment such as goggles, gloves and overalls.

✶ Change cylinders away from sources of ignition.

Oxygen can cause explosions, if used with incompatible materials. In particular, oxygen reacts explosively with oil and grease. Therefore:

* Never allow oil or grease to come into contact with oxygen valves / cylinder fittings.
* Never use oxygen with equipment not designed for it.
* In particular, check that the regulator is safe for oxygen and for the cylinder pressure.

Applicable Legislation / Standards

* SHWW (General Application) Regulations, 1993, Part IV: Work Equipment Regulations.
* SHWW (Construction) Regulations, 2001.
* SHWW (Explosive Atmospheres) Regulations, 2003.
* SHWW Act, 2005.
* *Proposed SHWW (Construction) Regulations, 2006.*
* *Proposed SHWW (General Application) Regulations, 2006.*

* IS / EN 470:1995. Protective Clothing for Use in Welding & Allied Processes - Part 1: General Requirements.

Further Information

* HSE: *Safety in Gas Welding, Cutting & Similar Processes*, INDG 297.
* HSE: *The Safe Use of Compressed Gases in Welding, Flame-cutting & Allied Processes*, HSG139.

WELFARE FACILITIES

See also: Cold / Fire / Lighting / Pregnant Employees / Site Security

What are welfare facilities?

Welfare includes general stability, energy distribution, emergency exits and plans, fire detection, fire equipment, weather protection, ventilation, temperature, lighting, doors, gates, traffic routes, loading bays, welfare facilities, sanitary conveniences, access and egress, facilities for pregnant and breast-feeding mothers, facilities for handicapped workers.

Hazards → Risks

Infection

Food poisoning / disease

→

Ill-health

Occupational diseases - dermatitis

Controls → Managing the Risks

* Locate welfare facilities on, or adjacent to, the site and size them according to the number of workers on site (main contractor's responsibility).
* Keep all facilities clean and tidy at all times. Implement a 'clean as you go' policy.
* Do not store materials or plant in the canteen, dry room or toilet facilities.

Changing area / Dry room
* Provide a dry room for shelter from bad weather and for storing wet clothing overnight.
* Provide lockers for the storage of PPE.
* Provide separate changing rooms for men and women.
* Ensure that a daily cleaning plan is in place.

Canteen facilities
* Ensure that canteens have adequate space / tables / chairs to accommodate workers.
* Ensure that canteens are clean and hygienic at all times.
* Provide facilities for boiling water and heating food.
* Provide an adequate supply of wholesome drinking water.
* Ensure that a daily cleaning plan is in place.

Sanitary conveniences
* Ensure that all toilets discharge into the main sewer, where possible.

* Keep all toilets clean and tidy and ensure that they are suitably ventilated at all times.
* Provide separate facilities for men and women.
* In men's facilities, provide one stall (not urinal) for every 20 persons on site.
* Do not connect toilet facilities with the canteen or any work area. Ensure that they are easily accessible.
* Ensure that a daily cleaning plan is in place.
* Provide the requisite toiletries.

Washing facilities
* Ensure the site has adequate washing facilities.
* For work extending over 30 days, provide wash basins and drying facilities.
* Provide hot and cold water for washing.
* For work extending over 12 months (or where there are more than 100 workers on site), provide six sink units and, for every 20 additional workers over 100, another sink unit.
* Provide shower facilities, depending on the nature of the work.

Applicable Legislation

* SHWW (General Application) Regulations, 1993, Reg.17; Second, Third & Fourth Schedules.
* SHWW (Construction) Regulations, 2001, Fourth Schedule.
* SHWW Act, 2005.
* *Proposed SHWW (Construction) Regulations, 2006.*
* *Proposed SHWW (General Application) Regulations, 2006.*

Further Information

* HSA: *Guidelines to the SHWW (Construction) Regulations, 2001.*
* HSE: *Provision of Welfare Facilities at Fixed Construction Sites*, CIS18.
* HSE: *Provision of Welfare Facilities at Transient Construction Sites*, CIS24.

WORKING AT HEIGHT

The biggest killer on-site in 2004 - and for the past 10 years

See also: Chains & Slings / Cranes / Fall Protection Equipment / Falling Objects / Forklifts / Ladders / Lifting Equipment & Operations / Mobile Elevating Working Platforms / Openings / Pre-cast Elements & Components / Roof Work / Scaffolding / Trestles

Note: **These are very general guidelines; see more specific hazards above for more Controls.**

Hazards → Risks

Falls from a height   →   **Death**
Falling objects   **Serious injury**

Controls → Managing the Risks

* Before permitting work at height, carry out a Risk Assessment.
* Ensure that there is a safe method of access and egress.
* Ensure that the work platform is the most suitable for the task, capable of supporting the intended weight and that it is secure.
* Ensure that the appropriate fall protection / PPE is in position / worn.
* Ensure that safety harnesses are kept in good condition and inspected regularly.
* Ensure that all workers have received training for work at height.
* Do not interfere with safety devices for work at heights.

Working at Height Risk Assessment:
1. **Avoid the risk:** Avoid all work at height.
2. **Reduce the risk:** Carry out as much work as possible at ground level.
3. **Collective fall protection:** Guard-rails, barriers, etc.
4. **Personal fall protection:** Fall restraint device.
5. **Collective protection:** Safety nets / soft-landing system.
6. **Personal fall arrest:** Harness / fall arrest lanyard.
Source: Des Brandon, NISO Eastern Region Lecture, *HSR*, June 2005, p.29.

Soft landing systems

The Soft Landing System has been designed for use principally inside a building during construction, where the bags will be enclosed by walls or partitions.

The bags are most commonly used at ground floor level while the joists are being boarded. This ensures that the carpenter will have a soft landing in the event of a leading edge fall while affixing the boards across the joists. On first or subsequent floors, the system may be used to protect workers while trusses are being installed, by installing bags on the covered joists.

Applicable Legislation

* SHWW (General Application) Regulations, 1993.
* SHWW (Construction) Regulations, 2001, Part 13: Working at Heights, Regs.51-79.
* EU Directive 2001/45/EC, Temporary Work at Height. (Requirements to be implemented in proposed new SHWW (General Application) Regulations, to be law in early 2006.)
* SHWW Act, 2005.
* *Proposed SHWW (Construction) Regulations, 2006.*
* *Proposed SHWW (General Application) Regulations, 2006.*

Further Information

* HSE: *Avoiding Falls from Vehicles*.
* HSE: *Height Safe - Absolutely Essential Health & Safety Information for People who Work at Height*.
* HSE: *The Work at Height Regulations 2005: A Brief Guide*, INDG401.
* NISO: *Construction Summary Sheets*.

YOUNG WORKERS / APPRENTICES

Hazards → Risks

Untrained workers
Inexperienced workers   **Injuries**
Incompetent workers   **Serious injuries**
Unsupervised activities   **Death**

Controls → Managing the Risks

* Do not employ persons under 16 on a construction site.
* Ensure all new staff are in possession of a current Safe Pass card.
* Ensure all new / young employees undergo Induction training before working on site. This Induction training should include the contents / implications of the Safety Statement, fire safety, general safety, the wearing and use of PPE, care and maintenance of PPE. Keep records of all such training.
* Do not allow risk-taking, horseplay and 'hazing' of young workers on site.
* Ensure that all young / inexperienced workers are under the direct supervision of a competent person for the duration of their training. Only allow them to work unsupervised when site management is satisfied that the person is competent.

Applicable Legislation

* SHWW (General Application) Regulations, 1993.
* Protection of Young Persons (Employment) Act, 1996.
* SHWW (Children & Young Persons) Act, 1997.
* SHWW Act, 2005.
* *Proposed SHWW (Construction) Regulations, 2006.*
* *Proposed SHWW (General Application) Regulations, 2006, Part X: Protection of Children & Young Persons.*

Further Information

* Department of Enterprise, Trade & Employment: *Protection of Young Persons (Employment) Act, 1996 - Guide for Employers and Employees.*

* FÁS - Apprentice Schemes.
* International Labour Organisation (ILO).

4: Construction Health & Safety Directory

4.1 Key Health & Safety contacts

Health & Safety Authority (HSA)
10 Hogan Place, Dublin 2
Tel: (01) 614 7000
Fax: (01) 614 7020
E-mail: infotel@hsa.ie
Web: www.hsa.ie

* ✶ Athlone Regional Office - (090) 649 2608
* ✶ Cork Regional Office - (021) 425 1212
* ✶ Galway Regional Office - (091) 563 985
* ✶ Limerick Regional Office - (061) 419 900
* ✶ Sligo Regional Office - (071) 914 3942
* ✶ Waterford Regional Office - (051) 875 892

Construction Industry Federation (CIF)
Construction House, Canal Road, Dublin 6
Tel: (01) 406 6000
Fax: (01) 496 6953
Web: www.cif.ie

* ✶ CIF Southern Region, Cork - (021) 435 1410
* ✶ CIF Western / Midland Region, Galway - (091) 502 680

4.2   Useful Websites

Abrasive Wheels
http://www.irishabrasives.ie   Irish Abrasives Ltd

Asbestos
http://www.arca.org.uk   Asbestos Removal Contractors
                                          Association

Chemicals
http://ecb.jrc.it/classification-labelling   European Chemicals Bureau -
                                          chemicals classified as carcinogens
                                          / mutagens / toxic to reproduction
http://www.rpii.ie   Irish Radiological Protection
                                          Institute of Ireland
http://mahbsrv.jrc.it   Major Accidents Hazards Bureau
                                          (Seveso)
http://www.msdssearch.com   MSDS Search

Concrete
http://www.irishconcrete.ie   Irish Concrete Federation

Construction
http://www.cif.ie   Construction Industry Federation
http://www.iei.ie   Institution of Engineers of Ireland
http://www.irishbuildingindustry.ie   Irish Building Industry Directory
http://www.riai.ie   Royal Institute of the Architects of
                                          Ireland

Cranes
http://www.irishcranes.com   Irish Crane & Lifting Limited

Demolition
http://www.iie-online.ie   Irish Industrial Explosives Ltd.

Electricity
http://www.esb.ie   ESB
http://www.reci.ie   Register of Electrical Contractors of
                                          Ireland

First Aid
http://www.civildefence.ie   Civil Defence School
http://www.firstaid.ie   First Aid Ltd - courses / training
http://www.irishheart.ie   Irish Heart Foundation

http://www.redcross.ie	Irish Red Cross Society
http://www.nats.ie	National Ambulance Training School
http://www.nifast.ie	Nifast
http://www.orderofmalta.ie	Order of Malta Ambulance Corps St John Ambulance Brigade of Ireland

Gas

http://www.bocgases.ie	BOC Gases
http://www.bordgais.ie	Bord Gáis Éireann

General Health & Safety

http://www.asce.org	American Society of Civil Engineers
http://www.asthmacare.ie	Asthma Care Ireland
http://www.asthmasociety.ie	Asthma Society
http://www.ladders-blma.co.uk	British Ladder Manufacturers Association
http://www.bsi-global.com	British Standards
http://www.cibse.org	Chartered Institution of Building Services Engineers
http://www.cif.ie	Construction Industry Research & Information Association
http://www.epa.ie	Environmental Protection Authority
http://europe.osha.eu.int	European Agency for Safety & Health at Work
http://www.europa.eu.int	European Commission Health & Safety at Work
http://www.fas.ie	FÁS
http://www.hsa.ie	Health & Safety Authority
http://www.hseni.gov.uk	Health & Safety Executive, Northern Ireland
http://www.hse.gov.uk	Health & Safety Executive, UK
http://www.healthandsafetyreview.ie	Health & Safety Review Magazine
http://hibernian.netsource.ie/ pdfs/gen_Driver_Handbook.doc	Hibernian Insurance - personalisable Drivers' Handbook
http://www.injuryprevention.org	Injury Prevention Website
http://www.iosh.co.uk	Institute of Occupational Safety Health
http://www.irishsafetycentre.com	Irish Safety Centre - a central forum for all Safety Advisers and consultants
http://www.publichealth.ie	Institute of Public Health Ireland

http://www.iarc.fr	International Agency for Research on Cancer
http://www-iea.me.tut.fi	International Ergonomics Association
http://www.cancer.ie	Irish Cancer Society
http://www.constructionworkers health.com	Irish Construction Workers Health Trust
http://www.iws.ie	Irish Water Safety
http://www.nsai.ie	National Standards Authority of Ireland
http://www.niso.ie	NISO
http://www.osha.gov	Occupational Safety & Health Administration
http://www.opw.ie	Office of Public Works
http://www.roofingcontractor.com	Roofing Contractor magazine
http://www.rospa.co.uk	Royal Society for the Prevention of Accidents
http://www.sii.ie	Security Institute of Ireland
http://www.who.int	World Health Organisation

Labour Relations / Employees

http://www.entemp.ie	Department of Enterprise, Trade & Employment
http://www.equality.ie	Equality Authority
http://www.ilo.org	International Labour Office
http://www.ibec.ie	Irish Business & Employers Confederation
http://www.lrc.ie	Labour Relations Commission

4.3 Professional Associations

(contact the CIF for contact details and further information)

Alliance of Specialists Contractors Associations

Architectural & Monumental Stone Association

Association of Professional Project Managers

Bricklaying & Allied Trades Contractors Association (BATCA)

Civil Engineering Contractors Association (CECA)

Concrete Manufacturers Association

Construction Surveyors Institute

Electrical Contractors Association (ECA)

Electrical Contractors Safety & Standards Association (ECSSA)

Environmental Services Contractors Association (ESCA)

Equipment Hire Association

Floor Covering and Tiling Contractors Association (FCTCA)

Formwork Contractors Association

Insulating Contractors Association

Irish Association of Demolition Contractors (IADC)

Irish Contractors Plant Association (ICPA)

Irish Home Builders Association (IHBA)

Irish Kitchen & Fitted Furniture Association (IKFFA)

Irish Platform Contractors Association

Irish Preservation & Damp Proofing Association (IPDPA)

Irish Property Developers Association

Irish Shopfitters' Association

Irish Steel Fixing Association

Irish Window Association (IWA)

Joinery Manufacturers Association (JMA)

Master Builders & Contractors Association (MBCA)

Master Glaziers Association (MGA)

Master Painters & Decorators of Ireland (MPDI)

Mechanical Engineering & Building Services Contractors Association (MEBSCA)

Mechanical Engineering Contractors' Association (MECA)

National Association of Scaffolding & Access Contractors (NASAC)

National Concrete Producers Association

National Furniture Manufacturers' Association (NFMA)

Plastering Contractors Association (PCA)

Plumbing & Heating Contractors Association

Register of Electrical Contractors of Ireland (RECI)

Road Marking Contractors Association (RMCA)

Roof Manufacturers & Suppliers Association (RMSA)

Roofing & Cladding Contractors Association (RCCA)

Royal Institute of Architects of Ireland

Sheetmetal Manufacturers Association (SMA)

Society of Chartered Surveyors

Structural Steel Manufacturers Association.

Other Health & Safety titles from
OAK TREE PRESS

THE IRISH HEALTH & SAFETY HANDBOOK
2nd edition
Thomas N Garavan
€110 hb : ISBN 1-86076-189-5

This revised and comprehensive second edition of The Irish Health & Safety Handbook covers all aspects of health & safety theory and practice in the working environment. It also focuses on the importance of proper health & safety measures in relation to the outside environment, and companies' responsibilities. It is an indispensable source of guidance for safety & health practitioners, HR specialists, industrial engineers, and students of all programmes in health, safety and the environment.

CIVIL LIABILITY FOR INDUSTRIAL ACCIDENTS
John PM White
€540 (3 Volume set) : ISBN 1-86076-198-4

This uniquely comprehensive work in three volumes provides an in-depth and up-to-date analysis of all aspects of employers' liability at common law and the statutory regime of protection of workers' safety and health. Extending to 4,300 pages, Dr White's work is an indispensable reference in the library of anyone who has a professional responsibility in relation to industrial accidents.

SURVIVING STRESS
A Guide for Managers & Employees
Sam Malone
€25 pb : ISBN 1-86076-295-6

Drawing from a wide range of recent research, and full of practical advice and invaluable information, Surviving Stress will prove invaluable to managers and those responsible for formulating strategies to combat stress in the workplace. It will also act as a source of ideas for trainers designing stress management programmes. Employees who want to take a more proactive approach to stress avoidance and stress management will find plenty of ideas in this book to help them achieve that end.

To order:
Email: orders@oaktreepress.com
Tel: 1890 313855 / 021 431 3855 Fax: 021 431 3496